D1101174

Three dimensional applications in Geographical Information Systems

Three dimensional applications in Geographical Information Systems

Edited by

Jonathan Raper

Department of Geography
Birkbeck College
University of London

Taylor & Francis
London New York Philadelphia
1989

USA   Taylor & Francis Inc., 1900 Frost Road, Suite 101, Bristol, PA 19007
UK   Taylor & Francis Ltd., 4 John St., London, WC1N 2ET

British Library Cataloguing in Publication Data
Three dimensional applications in Geographical Information Systems.
    1.Geophysics. Technique
    I. Raper, Jonathan
    551'.028

    ISBN 0-85066-776-3

Library of Congress Cataloging-in-Publication Data
  is available

Cover Design by Jordan and Jordan, Fareham, Hants, UK
Cover diagram produced by Interactive Volume Modeling software from
Dynamic Graphics Inc.

Printed by Burgess Science Press, Basingstoke, Hampshire, UK

Contents

Preface   vii

Contributors   ix

1.   Three dimensional display of geologic data
J. Nicholas Van Driel   1

2.   The 3-dimensional geoscientific mapping and modelling system: a conceptual design
Jonathan F. Raper   11

3.   Surface interpolation, spatial adjacency and GIS
Christopher M. Gold   21

4.   Efficient mapping of heavy metal pollution on floodplains by co-kriging from elevation data
Henk Leenaers, Peter A. Burrough and Joop P. Okx   37

5.   The application of a digital relief model to landform analysis in geomorphology
Richard Dikau   51

6.   Visualisation of digital terrain models: techniques and applications
Robin A. McLaren and Tom J. M. Kennie   79

7.   Computer-assisted cartographical 3D imaging techniques
Menno J. Kraak   99

8.   The role of three-dimensional geographic information systems in subsurface characterization for hydrogeological applications
A. Keith Turner   115

9.   Spatial data structures for modeling subsurface features
Carl Youngmann   129

10.   Creating a 3-dimensional transect of the earth's crust from craton to ocean basin across the N. Appalachian Orogen.
John D. Unger, Lee M. Liberty, Jeffrey D. Phillips, and Bruce E.Wright   137

vi

11.   Three-dimensional GIS for the earth sciences
Dennis R. Smith and Arthur R. Paradis   149

12.   Three dimensional representation in a Geoscientific Resource Management
System for the minerals industry
Peter R. G. Bak and Andrew J. B. Mill   155

Index   183

Preface

The study of process, form and spatial interrelationships in the geosciences must be carried out, axiomatically, in 3 dimensions. The 3 dimensional computer modelling tools available to the geoscientist in the 1970s and early 1980s only developed slowly along with computer technology in the last decade, as typical tasks in the geosciences are computationally very demanding and the data complex in character. Recently however, developments in computer graphics, spatial theory and a marked improvement in the price/ performance ratio of hardware have stimulated an explosive growth in software suitable for use in geoscientific modelling. This new environment of growth appears to have initiated a self-perpetuating technological cycle: faster hardware reduces the time taken on a task; more complex tasks are programmed leading to a demand for more processing power; and so the demand for yet faster hardware progresses. Geoscientists in a wide variety of sectors are now engaged in developing and using this new generation of tools: that is the environment for the production of this book.

My reasons for bringing together the contributions to this book are twofold. Firstly, it became clear to me recently when conducting a survey of 3D mapping and modelling software for the British Geological Survey that many researchers are working in this field in parallel around the world, who would clearly benefit from an exchange of ideas. Secondly, for those who are entering this field there is currently no up to date survey of research, and no accessible discussion of the basic design principles for geoscientific mapping and modelling systems. I hope that this book will meet some of these needs.

The book is broken loosely into 5 sections forming a progression from design to implementation. The first section contains two contrasting papers which introduce 3D GIS. Firstly, *Nick Van Driel* presents an overview of the available tools and requirements for 3 dimensional display, and discusses some of the applications of these tools at the USGS. Secondly *Jonathan Raper* presents a conceptual design for the development of a geoscientific mapping and modelling system, considering the relative merits of different data structures and identifying the key aspects of query design for geoscientific data.

The second section concerns the use of surfaces to model geoscientific data. *Chris Gold* discusses various approaches to contouring scattered data sets and argues for the use of adjacency rather than distance criteria in interpolation. This contrasts with the paper by *Henk Leenaers, Peter Burrough and Joop Okx* who consider how co-kriging of elevation with lead contamination values can lead to better descriptions of contamination distribution than standard surface fitting algorithms. *Richard Dikau* shows how surface modelling can be applied to the problem of automatic landform classification from publicly available digital elevation models.

The third group of papers consider the techniques and problems of visualisation as applied to geoscientific data. *Robin McLaren and Tom Kennie* discuss the range of techniques for rendering images and illustrate some typical applications, whilst *Menno Kraak* considers the problems of cartographic design in 3D from the point of view of perception.

A fourth set of papers contain some application studies of 3D GIS in use, and define many of the practical problems which arise in modelling. *Keith Turner* considers the application of 3D GIS to dynamic modelling in hydrogeology and shows a diagram suggesting how information should flow through a modelling system. This is followed by a paper by *Carl Youngmann* who considers the data structure alternatives for the handling of geo-objects in 3D from experience in the petroleum exploration industry. Thirdly, *John Unger et al.* show how 3D GIS is in use to its practical limits in the geophysical modelling of the earth's crust in Central Maine.

The last two papers describe functioning 3D GIS systems. *Dennis Smith and Art Paradis* describe the facilities available in IVM –a new 3D GIS from Dynamic Graphics– and illustrate the operation of a number of the spatial functions in a set of colour plates. Finally, *Peter Bak and Andrew Mill* review the CAD literature, and describe the implementation of a 3D GIS using octrees as a spatial data structure.

The papers contained in this book were presented at conferences held in the autumn of 1988 and the Spring of 1989 and so represent at the time of writing the most up to date collection of published works in this field. It should be noted that the book is a collection of disparate work gathered together for this book and was not commissioned specifically: as such there may be some gaps in the coverage of this new field. These are for future authors to fill!

Finally, no preface would be complete without thanks to those who have made this book possible. Thanks firstly to the authors who all submitted discs or electronic mail submissions: this book has been produced using Microsoft Word™ 3.01 on an Apple Macintosh II in under 2 months and output in camera ready form. Many thanks to Apple UK for the loan of the Mac II for this and other projects at Birkbeck.

However, thanks above all to my wife Frances who has suffered most from the constant demands of "the book". Without her support, the project would never have been finished!

Jonathan Raper
Birkbeck College
March 1989

Contributors

Peter R. G. Bak   Computer Aided Design Research Group, Dept. of Mineral Resources Engineering, Royal School of Mines, Imperial College, London SW7 2BP, UK.

Peter A. Burrough   Department of Geography, State University of Utrecht P.O. Box 80.115, 3508 TC Utrecht, The Netherlands.

Richard Dikau   Department of Geography, University of Heidelberg, Im Neuenheimer Feld, 348 D-6900 Heidelberg, Federal Republic of Germany.

Christopher M. Gold   Department of Geography, Memorial University of Newfoundland, St. John's, Newfoundland, A1B 3X9, Canada.

Tom J. M. Kennie   Balfour Beattie Ltd., 7 Mayday Road, Thornton Heath, London, UK

Menno J. Kraak   Department of Geodesy, Delft University of Technology, Thijsseweg 11, 2622 JA Delft, The Netherlands

Henk Leenaers   Department of Geography, State University of Utrecht P.O. Box 80.115, 3508 TC Utrecht, The Netherlands.

Lee M. Liberty   United States Geological Survey, 922 National Centre, Reston, VA 22092, USA

Robin.A. McLaren   Know Edge Ltd., 33 Lockharton Avenue, Edinburgh, EH14 1AY, UK.

Andrew J. B. Mill   Computer Aided Design Research Group, Dept. of Mineral Resources Engineering, Royal School of Mines, Imperial College, London SW7 2BP

Joop P. Okx   BKH Consulting Engineers, P.O. Box 93224, 2509 AE The Hague, The Netherlands.

Arthur R. Paradis Dynamic Graphics Inc., 2855 Telegraph Avenue, Berkeley, California 94705, USA

Jeffrey D. Phillips United States Geological Survey, 922 National Centre, Reston, VA 22092, USA

Jonathan F. Raper Dept. of Geography, Birkbeck College, 7-15 Gresse St., London W1P 1PA, UK

Dennis R. Smith Dynamic Graphics Inc., 7201 Wisconsin Avenue, Suite 640, Bethesda, MD 20814, USA

A. Keith Turner Dept. of Geology and Geological Engineering, Colorado School of Mines, Golden, Colorado 80401, USA

John D. Unger United States Geological Survey, 922 National Centre, Reston, VA 22092, USA

J. Nicholas Van Driel United States Geological Survey, 904 National Centre, Reston, VA 22092, USA

Bruce E.Wright United States Geological Survey, 521 National Centre, Reston, VA 22092, USA

Carl Youngmann Sierra Geophysics Inc., 11255 Kirkland Way, Kirkland, WA 98033, USA

Chapter 1

Three dimensional display of geologic data

J. Nicholas Van Driel

Introduction

The increasing volume of available data and developments in computer graphics have provided an environment in which automated techniques for analysis and display of geologic information in three dimensions can be used to solve complex geologic problems. Computer techniques that are used to study geology can be divided into three classes: two dimensional analysis and display, three dimensional display, and three dimensional analysis. 2D techniques have been used by geologists for many years in the form of software for computer graphics, image processing, and more recently, as geographic information systems or GIS. 2D display representations include surfaces, multiple layers, fence diagrams, and stereo images. These 2D processes are excellent tools for storing, manipulating and combining surfaces, but geologists require 3D capabilities for most applications.

The next step up in complexity, 3D display, provides better ways to examine and communicate information through the use of fence diagrams, isometric surfaces, multiple surfaces, and stereo block diagrams. However, most 3D display techniques are limited to the two dimensional format of a CRT screen or plotter paper. The displays can be viewed, but they can't be analyzed as discrete entities, for example the solid cannot be measured, stretched, reshaped, or combined. True 3D analysis or 3D GIS, with a continuous volumetric data structure and appropriate analytical functions, would give geologists the tool to integrate a variety of data sources, store all the available information about a portion of the earth's crust, and operate on solid bodies as discrete entities. The further capacity to manipulate and analyze a 3-D body is the next logical technique that should be made available to geologists.

This paper focuses on the middle range of capabilities, 3D display, and describes some of the techniques that are available to geologists to help them analyze and communicate their information. The capabilities are available on a variety of machines, and use both proprietary and user-written software. The paper concludes with a discussion of the general requirements for 3D analysis.

Three-dimensional display

The most important advantage of using 3D displays is the way they appeal to our brains and to our eyes. A 2D plot of individual elevations on a surface doesn't spark much of an image when we look at it; a contoured surface is a little better, but the viewer is required to build the image in his mind. A wireframe perspective display in color makes the surface come alive. All of the details, as well as the general trends, are immediately visible. 3D displays portray data, which are a sample of the real world, in a manner that resembles how they actually appear in the real world. It is estimated that fifty percent of the brain's neurons are involved in vision. 3D displays light up more neurons and thus involve a larger portion of our brains in solving a problem. This phenomenon is particularly important to geologists who use visualization to solve 3D problems by observing a relatively small portion of the rock units that are exposed, and then using judgment, experience, and imagination to extrapolate the areas that are hidden.

The computer-aided 3D display techniques mimic the manual techniques that have been developed and used by geologists for many years. Contours on a surface from point elevations, fence diagrams, multiple surfaces and perspective drawings, originally done by hand, can now be produced quickly and efficiently by a wide variety of computer programs.

Surfaces

Surfaces are used in a variety of ways to represent geological information. The most common form, structure contour maps, can communicate the shape of a surface to people who are trained to use contour maps, but it takes some time and visualization skill to organize the information and build a mental picture of the surface. In contrast, anyone can instantly see the shape and details of a surface that is displayed as a wireframe perspective. The viewer of the 3D display can spend less time creating the image and more time analyzing it. Figure 1.1 shows the difference in these two views.

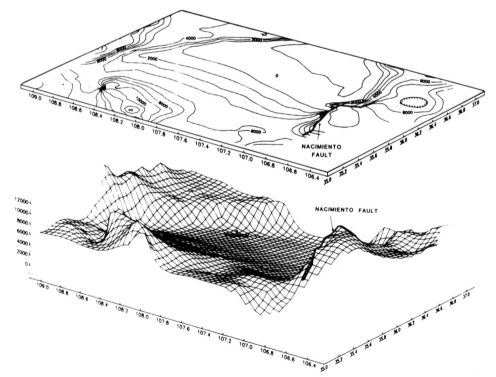

Figure 1.1   (a) Structure contour map on the base of the Dakota Sandstone, San Juan Basin (above)
(b) Wireframe perspective view of the same data (below)

Highly faulted surfaces are very difficult to visualize without the help of a 3D display. In addition to providing a more realistic display, automated 3D techniques allow the user to rotate and change the viewing angle to see all aspects of the subject. The Interactive Surface Modelling (ISM) software produced by Dynamic Graphics produced the complex rendition of subsurface faulting in Figure 1.2. Without a graphic like this, it would be very difficult to understand the nature of this faulted surface.

BASE OF BASAL SERIES

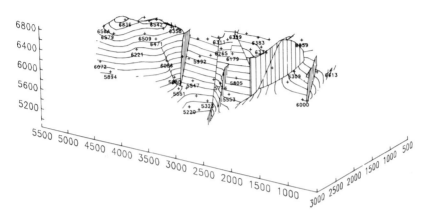

Figure 1.2 Structure contour map of a highly faulted surface produced by Interactive Surface Modelling software.

Combining data sets by draping one over the other is a powerful way to examine the relationships between the two. The most common uses of this technique combine surface and subsurface geologic information with a digital elevation model (DEM). For example, a bedrock geologic map draped over the topography will show the relationships between lithology and landforms, while a combination of a DEM and gravity map will display the local effects of topography on the gravity measurements.

Layers

A series of vertically registered surfaces can be used to represent a portion of the earth's crust by presenting information on individual subsurface layers. The intervening space can be "filled in" by the viewer. This technique is popular because it displays data in the form often available to geologists, discrete surfaces rather than continuous volumetric data.

The maps in Figure 1.3 show horizontal slices at 165m intervals through Cambrian rocks in the Springfield, Missouri USGS map sheet quadrangle. Sixteen major lithofacies were identified and mapped as layers using drill core and water well records. The surface area of each map is approximately 100 km by 200 km. The dark areas represent dolomite formed by destructive recrystallization of limestone. The stacked view in Figure 1.4 can be rotated and viewed from any angle or elevation. These displays permit the viewer to see relationships in 3D that are not evident from the source maps. In this example, geologists were able to examine the patterns of recrystallization, and to apply models of depositional environments to understand the depositional history of the region.

The layer technique has been used with excellent results for illustrations in several of the other papers in this volume, including those by Unger et al.

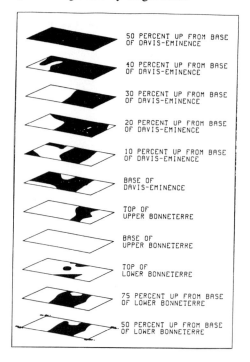

Figure 1.3 Layer maps showing the distribution of dolomite (black) and limestone (white) at 165m increments of depth.

Fence Diagrams

The fence diagram is a classic technique for displaying geologic information in three dimensions. A fence diagram generally consists of boreholes or outcrops connected by cross-sections and presents a display that can be both intuitively appealing and rich in information. In the past, the use of fence diagrams was somewhat limited because of the daunting volume of work connected with their construction. With the computer programs that can now produce fence diagrams, this important display technique is more readily available to geologists.

The examples in Figure 1.5 show the distribution of five geologic units in the San Juan Basin in New Mexico. The viewer can visualize the shapes of these units because the profiles show cross-sections in two axes, and the viewer "connects" the profiles in his mind. Fence diagrams can be especially useful for projects with sparse data points and the requirement to cover large geographic areas.

Stereo

Geologists have used stereo views provided by airphotos for many years because stereo provides an excellent representation of the earth's surface. In many cases, the stereo view is better for mapping because of the vertical exaggeration inherent in the photos. Seismologists in the USGS have developed computer programs to calculate and plot stereo images of earthquake events in the subsurface.

The stereo plots in Figure 1.6 are from the Morgan Hill, California magnitude 6.2 earthquake of 1984. Each cube in the plots is 12 km on a side, with 2 km depth intervals marked on the side. The smaller internal cube is one kilometer on a side and is used for orienting the view. The main earthquake shock is plotted as a circle, and the plusses indicate the aftershocks. The viewer is observing the plot from a depth beneath the surface of 5 km. When these plots are viewed in stereo, the depth of view is excellent. The fault plane is well

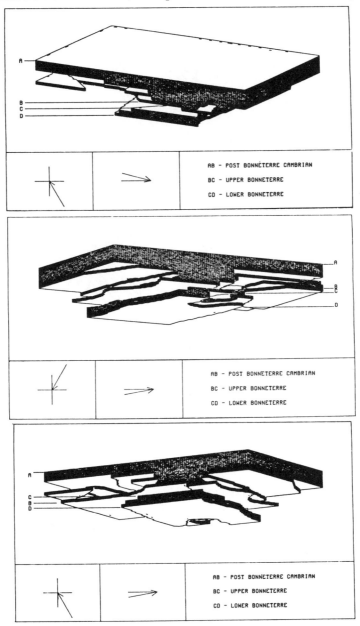

Figure 1.4 Stacked layers from Figure 1.3 produced by JKMAP software, viewed at an angle from (a) the SE, above the surface; (b) the NE below the surface; (c) the SE below the surface.

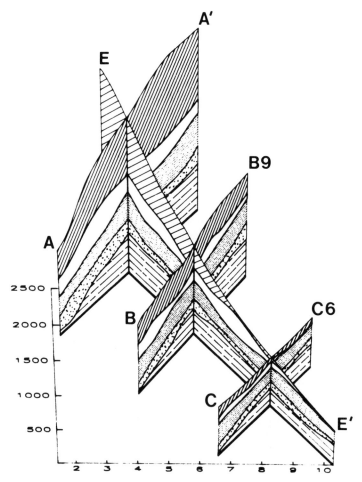

Figure 1.5 Fence diagram produced by Interactive Surface Modelling software.

defined, and can be seen receding into the distance. These plots were made on a Megatek Whizzard system, and show subsurface data in a perspective stereo view.

Hologram

The hologram consists of a thin plastic layer with a molded interference pattern that is covered by another plastic layer. The aluminum acts like a mirror and reflects white light waves through the interference pattern to create the three-dimensional image. This "rainbow hologram" technique has been used as a decorative addition to bank cards, and for illustrations in magazines. USGS seismologists have used rainbow holography to produce a three dimensional image of a 20 degree by 20 degree portion of the earth's crust showing seismic activity in the Aleutian Islands. The hologram shows the islands on the surface, a three-dimensional grid, and earthquake locations in the subsurface. Although this can be an effective means for communicating 3-D in a printed form, its use is limited because of the expense.

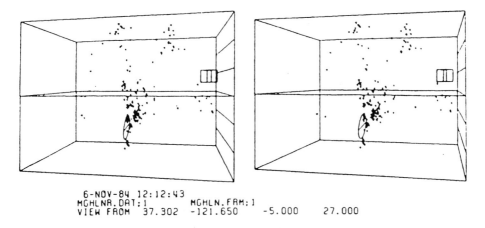

```
    6-NOV-84 12:12:43
    MGHLNR.DAT:1        MGHLN.FRM:1
    VIEW FROM  37.302  -121.650    -5.000    27.000
```

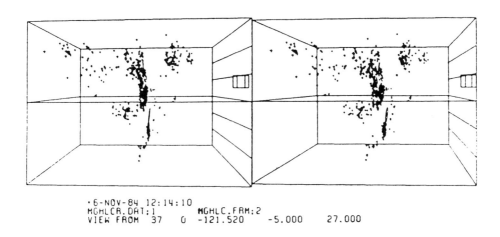

```
   •6-NOV-84 12:14:10
    MGHLCR.DAT:1        MGHLC.FRM:2
    VIEW FROM  37   0  -121.520    -5.000    27.000
```

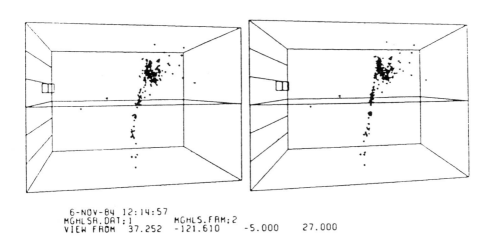

```
    6-NOV-84 12:14:57
    MGHLSR.DAT:1        MGHLS.FRM:2
    VIEW FROM  37.252  -121.610   -5.000    27.000
```

Figure 1.6 Stereographic views of the Morgan Hill earthquake showing aftershocks plotted in the subsurface. The three views are the north (top), central (middle) and south (bottom) sections of the fault.

J. Nicholas Van Driel

Surface Rendering vs Volume Rendering

The Pixar Image Computer is capable of producing three-dimensional images that show both the surface of an object, and internal details. The surface rendering technique is the equivalent of a hand-drawn block diagram with a map on the top surface and cross-sections on the faces of the block. The volume rendering technique, which uses a semi- transparent depiction of both surface and internal features, has no manual equivalent. The volume rendering technique is extremely powerful because it shows all the surface and subsurface details in their correct spatial relationships. Potential drawbacks to the use of this technique are the requirement of a relatively complete data set and expense of hardware and software.

Essential elements for 3D analysis

Three dimensional displays are useful for many applications, but many geologists want to use 3D analysis. The four essential elements required for true 3D analysis are:
1) subsurface data;
2) a 3D data structure;
3) hardware and software systems; and
4) geologists willing to learn new techniques.
Several of the papers in this volume address the design of a data structure and propose systems to accomplish 3D analysis, the data requirement is equally important.

Building a 3D data set for a portion of the earth's crust is very different from constructing a model for medical imaging or computer-aided design. Geologists are not able to acquire complete, continuous information. The information sources available to geologists include direct observation of outcrops, mines or cores, and indirect geophysical measurements. Each of these sources provides information for only a point, a profile, or a surface. Using these samples or pieces of the picture, the geologist must extrapolate and interpolate to construct a complete subsurface model. The quality and accuracy of 3D geologic models is severely constrained by the availability of subsurface information. In many cases, the lack of information may preclude the need for a sophisticated analytical capability.

Most geologists as computer users would like high performance with no requirement for programming. Some system developers have made their system interfaces extremely easy to use, while others require learning and practice by the user. Unfortunately, the good interfaces are not always developed for the new, state-of-the-art systems. The geologist is the analyst who uses computer tools, judgement, experience and imagination to manipulate the samples of data to build a 3-D model. Thus, the successful development of true 3D analysis depends on the cooperation of geologists who can perceive the benefits to be derived from an investment of time and effort in learning a new tool. System developers should consider that the user interface is as important as the analytical capability.

Conclusion

Many computer programs are available to geologists to help them analyze and portray geologic information. Geologists who use 3D display techniques have a clear advantage in their ability to visually analyze complex spatial data, and to communicate their concepts to colleagues and to non-geologists. The process of selecting a display technique involves a complex mixture of factors including the quality and quantity of subsurface information available, the analytical requirements of the project, the time available, the investigator's computing skill and inclination to learning new techniques, and finally, the cost and availability of the hardware and software. The wide availability of computer display techniques on both large and small machines makes it possible for nearly everyone to employ some form of 3D display capability.

Note

The use of trade names is for identification purposes only and does not constitute endorsement by the U.S. Geological Survey.

Chapter 2

The 3-dimensional geoscientific mapping and modelling system: a conceptual design

Jonathan F. Raper

Introduction

Following the considerable growth in geographical information systems to meet 2D mapping needs (Rhind 1987), attention is now turning to the design and construction of 3D geoscientific mapping and modelling systems (GMMS) in a range of application areas. Recent developments have been reported for example in the fields of geology (Vinken 1986), oil exploration (Ritsema et al. 1988), mining (Kavouras and Masry 1987) and civil engineering (Raper and Wainwright 1987). With the simultaneous improvement in the price/performance ratio for hardware and the rapid development of software for graphic modelling, increasing numbers of geoscientific modelling problems have begun to seem tractable.

Automated digital data capture and the demand for geoscientific information on-line means that large quantities of data are now accumulating in geoscience databases (Maurenbrecher & Kooter 1988) which both require efficient management <u>and</u> are subject to growing demands for sophisticated query response or 3D spatial modelling (Adlam, Clayton and Kelk 1988). However, whilst the visualisation tools have become progressively more sophisticated (Salmon and Slater 1987), less attention has been paid to the development of information processing tools for an integrated GMMS. This paper is a first attempt at presenting some design principles for such a development: the approach is highly data dependent, reflecting the common qualitative constraints on geoscience data, and concentrates on the selection of the appropriate spatial identity of a geo-object to be displayed by the visualisation software. The specification of a framework for a GMMS considered here is device independent at this stage given the current rate of technological change: micro-, mini- and mainframe computer solutions can all be envisaged.

Geo-object identity: data models for geoscientific data

An important distinction between GMMS in 2D and 3D space lies in the spatial identity of the geo-object under investigation. In the mapping and modelling of geo-objects in 3D space, particularly in the subsurface, the 3D domain is rarely known with any great confidence. This constraint reduces the quality of information available to the status of samples of a substantially unknown geo-object. It is necessary therefore to form a hypotheses about the geo-object first by defining its identity in 3D.

11

The use of data models to determine the 3D geo-object 'hypothesis'

An important qualitative difference is seen in the process of 3D spatial modelling between the identification of geo-objects which are believed to have a discrete spatial identity e.g. a perched aquifer or fault block, and those which vary in identity in space but which can be visualised by choosing threshold parameter values for inspection, e.g. ore grades or a sedimentary facies. 3D spatial modelling of the first kind of geo-object is *sampling limited*, and a minimum of database operations is required to assemble the spatial data, which will consist of a boundary type definition. However, the 3D modelling of the second type of geo-object is *definition limited* and is governed by the establishment of the data model used to establish a database query.

With the relative paucity of information often available to carry out such *definition limited* 3D spatial modelling, the importance of the data model used in the analysis increases. This is because the geo-object itself is usually defined by the sampling or selection of parameters established by the data model. Green and Rhind (1986) defined a geoscience data model as:

> "an abstraction of the real world...which incorporates only those properties
> thought to be relevant to the task(s) in hand"

The data model for a stratum believed to exist in the subsurface may be defined by lithological and structural parameters in a specific combination. The 3D spatial identity of the geo-object is then established by searching the population of selected characteristics for the boundaries of the defining conditions and recording the X,Y,Z coordinates. Note therefore that by altering the contents of the data model, and iterating the search process, a new set of X,Y,Z coordinates defining the object can be created. In the case of a study attempting to establish the overall architecture of a sedimentary sequence the basic spatial arrangement of geo-objects can often be defined in different ways (they may even overlap!), depending on the contents of the data model for each element of the sequence defined (Raper 1988).

It is clear therefore, that establishing the spatial identity of *definition limited* geo-objects in the subsurface is highly sensitive to the contents of the data model. The essential point is that the errors or bias inherent in the process of defining this data model can be as great, if not greater than those introduced in the sampling of the parameters defining the geo-object or in the process of its visualisation. In establishing the design principles for a GMMS it is recommended that this step be formalised as an element of "good practice".

Finally, it may be necessary to edit the data to select the values to be used in the 3D spatial modelling. For example it may be necessary to parse or validate the raw data, and subsequently to parameterise or regionalise the values before further analysis.

Sampling subsurface phenomena

Once a data model has been defined in terms of a parameter set or boundary type, the next step is to collect appropriate measurements in the study domain to determine the locations meeting these criteria. These locations may be in the form of X,Y,Z coordinates or 3D vectors and are then used to establish the full 3D spatial form of the geo-object. The collection of these measurements is, however, a sampling exercise and in the subsequent modelling these measurements should be considered as the spatial representation of the probability that a geo-object exists in a particular form.

Since a geo-object is likely to have an unknown variability in form, neither the correct sampling frame (random or structured) nor the sampling density can be determined in advance. Two main types approaches to this problem are evident in current practice. The first type of sampling adopted is governed by economic or project rather than geoscientific criteria. For example the sampling of a ground conditions for a new road is often constrained to the centreline of the carriageway. A typical procedure in this situation is to carry out an initial sampling operation when the variability of the 3D domain can be assessed, before following up with a structured sampling operation to investigate the areas of greatest variability or importance. An alternative procedure is to use a pre-planned sampling operation using a large scale (regular?) sampling frame or empirical knowledge to select sampling locations. This latter approach can utilise "soft data" about the geo-object (Schaeben 1989) such as regional dip trends in a stratigraphic unit, or covariance techniques such as co-kriging to identify potentially important locations (Leenaers, Burrough and Okx, this volume).

Another major difficulty in the sampling of 3D geo-objects is the occurrence of multi-valued surfaces. Many sedimentary structures such as overthrust folds, reverse faulted strata or perched acquifers generate multiple values of Z for the boundaries of the geo-object, which can not be represented by 2.5D surface modelling (Rüber 1989). This problem illustrates several drawbacks associated with point sampling of a geo-object using one dimensional vertical sampling lines:

- Firstly it can be very difficult to distinguish between the multiple occurrence of a similar object down a line and the repeated occurrence of a single object through folding or faulting;
- Secondly a vertical sample may not be orthogonal (as is often assumed) to the axis of primary object formation as in the case of a vertical pipe or contaminant plume; and
- Thirdly the continuity between any two occurrences of the boundary of a geo-object taken from two neighbouring point samples cannot be assumed to be straightforward.

A final sampling difficulty associated with 3D geo-objects arises in the case of dynamic phenomena such as acquifers and oil reservoirs, or in evolutionary systems such as coastal accumulation environments. It may be necessary to sample these data sets repeatedly over time which will generate multiple values of Z for a given X,Y position. These values may change over time according to positive process feedback indicating the operation of a local factor eg the greater accumulations on a coastal salt marsh near a major creek, or the drawdown of a piezometric surface near a well, providing further sampling difficulties.

Conceptual design of a Geoscientific Mapping and Modelling System

Following the successful definition of the spatial identity of a geo-object of interest, the data can be modeled within the framework of a GMMS. A variety of choices about data storage, graphical data structures and visualisation, however, constrain the scope of the modelling by defining the 3D functionality available. The sequence of operations is conceptually shown in Figure 2.1.

The structuring of 3D spatial data

The first option to consider in the design of a GMMS is the status and storage of the data available for 3D spatial modelling. Many applications of a GMMS will apply a data model to define the spatial data describing a geo-object, and accordingly produce meta-data i.e. data about data. If the meta-data is being used in visualisations then it is necessary to store this information either locally in temporary filespace, or ideally to maintain it permanently as a record of a particular hypothesis about the data. Geological maps, for example, may in future be interactively defined by the user and visualised in real time from a database: however, it is important for this definition to be stated. Even if simple X,Y,Z coordinates have been selected where the geo-object is clearly defined, the database query used to find the spatial data may be a valuable item of meta-data which may be stored in a query library or a knowledge base.

GEOSCIENTIFIC MAPPING AND MODELLING SYSTEM: CONCEPTUAL DESIGN

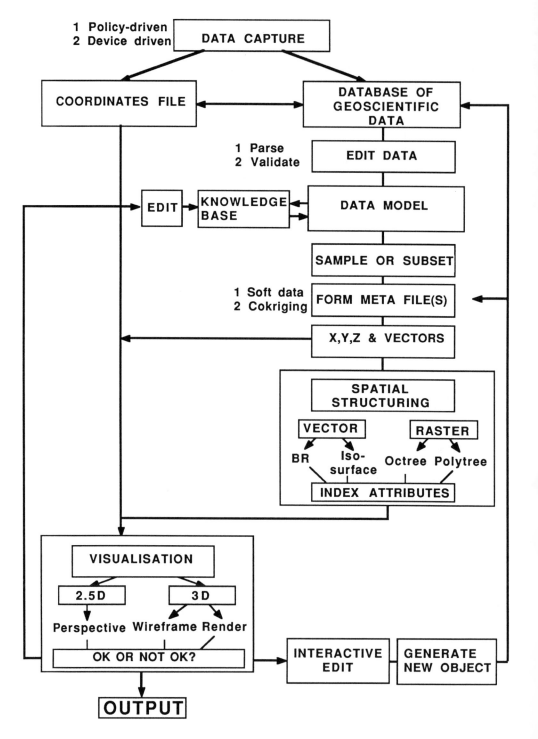

Figure 2.1 Conceptual design for the sequence of operations in a GMMS

Secondly, realising the optimum spatial functionality of a 3D modelling system may rely on the full 3D spatial structuring of the geo-object. It is currently common for the coordinates of a geo-object to be determined each time a model is formed since 3D spatial structuring for geoscience data has not hitherto been possible, and non-solid representations are difficult to query further. However, once 3D spatial structuring becomes common, the question of the recomputation of a model or the 3D spatial storage will arise. As processing power becomes cheaper per unit, the choice between the alternatives of recomputation and full 3D structuring of the data will be determined by the application involved and the storage space available.

The approaches to full 3D spatial structuring of a geo-object can be categorised into the 3D equivalents of raster and vector forms. The raster solutions to 3D data structuring mostly use spatial indexing systems and are based around the subdivision of the 'universe' of 3D space containing the geo-object into volume elements ('voxels'). The simplest form of storage for this data is as a binary raster, with each voxel stored as "on" or "off" depending whether the voxel contains a part of any object described in the scheme (Jones 1988). The indexing can be by 3D run encoding (Mark and Cebrian 1986) where all the voxels are visited by, for example the Morton Order, although this technique can develop huge demands for storage. A more sophisticated technique is the 3D equivalent of the quadtree called the octree (see Bak and Mill, this volume) which recursively divides space into 8 until any part of the subdivision is empty (outside the object) or full (inside the object), with the process continuing to a predetermined level of resolution. The advantage of this form of structuring is the very efficient conduct of Boolean operations on geo-objects.

However, a modified version of the octree called the polytree has recently been proposed by Carlbom (1987) which identifies the logical content of each voxel, i.e. whether full or empty, or a vertex, edge or surface cell of the geo-object. This scheme has the advantage of a solid representation amenable to rapid Boolean operations for 3D spatial operations, whilst also identifying the nature of the individual voxel, giving a pseudo object representation.

In the field of vector data storage for geo-objects there are two main forms of 3D spatial data structuring. The most common systems use topological relations to define 3D boundary representations for the indexing of geometrical data, such as the system of "simplicial complexes" described by Carlson (1987) or the "structured vector fields" described by Burns (1988). Attribute data can then be linked to this using an appropriate geo-relational system, although the processing overheads can be high. The alternative procedure is based on the 3D definition of iso-surfaces by 3D interpolation between points; Smith and Paradis (this volume) describe a new system of this type (IVM) based on the 3D triangulation of iso-surfaces, which are then shaded for a given light source for optimum visualisation.

Another strategy for 3D vector structuring is the spatial clustering of the vectors defining the geo-object by a geometrical attribute in a geoscience database. Work by Schek and Waterfeld (1986) and Horn et al. (1989) has led to the establishment of a prototype database (the "Geokernal") which implements the storage of 3D vector data items which are fully integrated with the associated attribute data. The storage structure used is based on the hierarchical subdivision of the primary data attributes into sub-objects, and by representing the 3D geometry within this structure as a set of points, polylines and cuboids spatially clustered by a regular or hierarchical subdivision of the 3D space enclosing the objects. The usual mode of access would be by specifying a 'clip' or 'compose' spatial query followed by a 'test' for the relevant non- spatial attributes. The chief advantage of this system is that it binds together the spatial and non- spatial components in a form of object-orientation, and allows the database to be set up for any data configuration. The main disadvantages are that it requires a complex database operation to set up each new data description and bind it to the database, and that it does not use an explicit data structure to store 3D geometry but subdivides space around the objects identified in each project.

The key to the efficient access to and use of 3D geoscientific data, however structured, are the spatial query routines. It is possible to define a set of generic 3D spatial query functions (analogous to those in 2D defined by Burrough 1986) which form a related set of spatial operations on 3D geo-objects: a first attempt at this classification is shown in Table 2.1 below:

SPATIAL FUNCTION	RASTER	VECTOR
AND	fast	slower
OR	fast	slower
XOR	fast	slower
NOT	fast	slower
TRANSLATE	slower	fast
ROTATE	slower	fast
SCALE	slower	fast
SHEAR	slower	fast
DISTANCE	fast	fast
ADJACENCY	fast	fast
SECTION (cut)	fast	slow
TUNNEL (bore)	fast	slow
BUILD (grow)	fast	slow
VOLUME	fast	slower
SURFACE AREA	fast	fast
CENTRE OF MASS	fast	slower
ORIENTATION	slow	fast

Table 2.1 Generic spatial functions in 3D spatial modelling and their effectiveness under different 3D data structures.

The table shows in outline terms how raster and vector spatial structuring affect the speed of operation of the 3D spatial functions. The relative merits of integral structuring and recomputation of the model also vary according to the type of data structure used. It is suggested that this form of analysis is the optimum way to decide on the type of spatial structuring which is appropriate for each set of data and operation.

Table 2.1 contains some functions which can be seen as enhancements of their 2D equivalents. These include the basic Boolean operations on the 3D entities modeled, the transformations using translation, rotation, scaling and warping, and adjacency and distance. However, in addition there are also functions which specifically concern solids, including the sectioning, tunneling and building of geo-objects. True 3D functions required of a GMMS may also include volume, surface area and centre of mass for a geo-object along with orientation of any defined axis. Very few if any of these "solid" functions have yet been fully implemented for real geoscience data sets with all their inherent complexity, and so only the theoretical performance of these queries is considered in the establishment of Table 2.1 (but see Bak and Mill, this volume).

Visualisation of 3-dimensional data

The primary subdivision of the available 3D visualisation tools is defined by the dimensionality of the visual representation used. At present 2.5D representations of surfaces drawn with perspective depth cues are common as part either of general purpose graphics libraries like GINO or specific geoscientific packages for terrain modelling such as ISM from Dynamic Graphics, GEOPAK from Uniras and the Intergraph DTM system. These systems when dedicated to geoscientific use often provide facilities to superimpose several surfaces which lie one above the other, giving an apparent reconstruction of a complex 3D structure. Systems running on mainframes or fast workstations with dedicated local hardware are capable of interactive rotation and the display of intersecting surfaces.

One of the reasons these 2.5D surface techniques have gained considerable acceptance in the geoscientific community lies in the normal thickness to area ratio of many of the geo-objects modeled. This ratio is usually extremely small reflecting the relative thinness of the objects relative to their wide extent, and is instrumental in making acceptable the representation of the geo-object by a plane. Note also that these applications use generalisations of the originally scattered data set to create the graphic models, computationally scaling the density of the grid of Z values generated, illustrating the smoothing inherent in the use of surfaces to explore the 3D form of subsurface geo-objects. It is axiomatic also, that the use of surfaces is not a holistic approach to modelling geo-objects: either the whole subsurface domain is not of interest in these cases, or the detailed architecture of the domain cannot be completely reconstructed: a surface is only one aspect of the full solid identity.

True 3D representations of geo-objects are normally generated from raw spatial data at present, and are often shaded to improve visibility of form. Many commercial systems exist to render this kind of visualisation (McLaren & Kennie, this volume), and with the publication of standards such as the "Renderman" protocol by Pixar Inc. (Pixar 1988) can be considered a well documented process. However, note that few 3D visualisations are generated directly from 3D spatial structures: until this is more common, these high quality rendered images will remain static images which cannot be interactively interrogated. One ultimate design objective in the development of a GMMS is the production of high quality images which can also be interrogated by graphic interaction with the model on the screen, perhaps using a 3D cursor such as the "Dataglove" recently demonstrated by the Media Lab at MIT (Aldersey-Williams 1989).

Conclusion

The foregoing discussion has emphasised the crucial role of the identity of the geo-object being modeled in any geoscientific application. The establishment, storage, editing and updating of the information in the data model that defines the geo-object is a vital issue for the design of the next generation of GMMS.

The process of visualising and analysing the geo-object of interest will become increasingly dominated by a need to choose an appropriate form of 3D spatial structuring. In the geosciences it is apparent that the ability to query the model will be at least as important as rendering the image in the most realistic way possible. Using a 3D spatial structure will permit a range of new spatial query functions: use of these new operators can be expected to generate new geo-objects of interest and extend the whole range of analysis.

These developments also point to new research opportunities: it will become possible and scientifically desirable to construct multiple models of the geo-objects under investigation and to analyse the spatial differences between different data models or forms of spatial structuring. The new capabilities provided by these new GMMS will, therefore, extend our ability to handle complex 3D objects whose identity is (unlike in CAD/CAM applications) established according to the probabilities associated with the data model and the sampling.

By selecting the important generic spatial functions from Table 2.1 and assessing the contents of the data model it is suggested that an appropriate 3D spatial structuring can be selected. The longer term future of this choice may, however, depend on the speed of the hardware and the access to the storage at- a- price: these market changes will also have a significant influence on the potential for development of geoscientific mapping and modelling systems.

References

Adlam, K. A. M. Clayton, A. R. and Kelk, B., 1988, A 'demonstrator' for the National Geosciences Data Index, *International Journal of Geographical Information Systems* **2**, 161-170.

Aldersey-Williams, H., 1989, A Bauhaus for the media age, *New Scientist* **1655**, 54-60.

Burns, K. L., 1988, Lithologic topology and structural vector fields applied to subsurface prediction in geology, *GIS/LIS '88, San Antonio, ACSM-ASPRS*.

Burrough, P. A., 1986, *Principles of geographical information systems for land resources assessment.* (Oxford: Oxford University Press).

Carlbom, I., 1987, An algorithm for geometric set operations using cellular subdivision techniques. *IEEE Computer Graphics and Applications*, **May 1987**, 45-55.

Carlson, E., 1987, Three dimensional conceptual modelling of subsurface structures. Chrisman, N. R. (ed), *Proceedings 8th International Symposium. on Computer Assisted Cartography, AutoCarto 8*, Baltimore, MD, pp 336-345,

Green, N. P. and Rhind, D. W., 1986, *Spatial data structures for geographic information systems*. Conceptual design of a geographic information system for the Natural Environment Research Council, Report **2**.

Horn, D et al., 1989, Spatial access paths and physical clustering in a low level geo-database system. *Geologisches Jahrbuch* **A 104** (Construction and display of geoscientific maps derived from databases).

Jones, T. A.,1988, Modeling geology in 3 dimensions, *Geobyte* **February 1988**, 14-20.

Kavouras, M. and Masry, S., 1987, An information system for geosciences: design considerations. Chrisman, N. R. (ed), *Proceedings 8th International Symposium. on Computer Assisted Cartography, AutoCarto 8*, Baltimore, MD, pp 336-345.

Leenaers, H., Burrough, P. A. and Okx, J., 1989, Efficient mapping of heavy metal pollution on floodplains by co-kriging from elevation data, (this volume).

Mark, D. M. and Cebrian, J. A., 1986, Octrees: a useful method for the processing of topographic and subsurface data. *Proceedings of ACSM-ASPRS Annual Convention* **1**, pp 104-113, Washington, D.C.

Maurenbrecher, P.M. and Kooter, B.M., 1988, The data jungle, *Delft Progress Report* **13**, 255-271.

McLaren, R. A. and Kennie, T., 1989, Visualisation of digital terrain models: techniques and applications (this volume).

Raper, J. F and Wainwright, D. E., 1987, The use of the geotechnical database GEOSHARE in site investigation data management. *Quarterly Journal of Engineering Geology* **20**, 221-230.

Raper,J. F., 1988, A methodology for the investigation of landform-sediment relationships in British glaciated valleys. Ph.D Thesis, Queen Mary College, Univ. of London.

Rhind, D. W.,1987, Recent developments in geographical information systems in the UK. *International Journal of Geographical Information Systems* **1**, 229-242.

Ritsema, I. L., Riepen, M., Ridder, J., and Paige, S.L., 1988, Global system definition for version 0.x of the GeoScientific Information System, TNO Institute of Applied Geoscience, SWS document: GSIS 1.1.

Rüber, O., 1989, Interactive design of faulted geological surfaces, *Geologisches Jahrbuch* A 104 (Construction and display of geoscientific maps derived from databases).

Salmon, R and Slater, M , 1987, *Computer Graphics*. (Reading, Mass.: Addison- Wesley).

Schaeben, H., 1989, Improving the geological significance of computed surfaces by CADG methods, *Geologisches Jahrbuch* A **104** (Construction and display of geoscientific maps derived from databases).

Schek, H.-J. and Waterfeld, W., 1986, A database kernal system for geoscientific applications, Marble, D. F. (ed),.*Proceedings of the Second International Symposium on spatial data handling*, Seattle, WA, pp 273-288.

Vinken, R.,1986, Digital geoscientific maps: a priority program of the German Society for the Advancement of Scientific Research. *Mathematical Geology* **18**, 237-246.

Chapter 3

Surface interpolation, spatial adjacency and GIS

Christopher M. Gold

Introduction

Spatial information may be viewed in two ways: as coordinates measured with a ruler along imaginary axes; or as the adjacency relationships between objects. The former has been the most common historically, but the latter appears more relevant to the handling of spatially distributed objects in a computer.

The interpolation problem, as implemented using weighted-average techniques, depends on the selection of appropriate neighbours to the sample location being estimated. Traditionally this was done using "metric" methods, with sometimes unsatisfactory results. An alternative "adjacency relationship" approach can be used, based on the construction of the Voronoi regions about each point, that removes many of the inconsistencies of the earlier methods and, in addition, provides an integrated weighting method for each adjacent point. Interpolants can then be modeled to provide appropriate levels of continuity.

The technique may be extended to data objects besides points, in particular to line segments, permitting consistent integration of elevation information at break-lines and contour segments as well as at point locations. While the concepts discussed here are applied to two spatial (independent) dimensions plus "elevation", the Voronoi method may readily be applied to higher dimensions.

Historical note

Since 1988 is the 300th anniversary of the "Glorious revolution" of 1688, I think it appropriate to review an argument that was going on in that era, between two proponents of the new scientific world view: Newton and Leibnitz. Somewhat over- simplified, the issue was whether space should be viewed as something tangible to be measured with a ruler or whether the only important information was the relationships between objects in that space.

This issue has surfaced many times since, not least in the development of GIS and automated mapping systems, but the particular topic I am concerned with here is also "ancient history" - interpolation, or contouring of scattered data points. Thus I would like to suggest that the "new" view encouraged by the current interest in "GIS" is the recognition that the spatial relationships between the objects on your map matter as much or more than their actual coordinates, particularly when using computers.

My original training was as a geologist. As with many colleagues, my first involvement with computers and maps involved the "contouring problem". This issue was a problem because it was considered somewhat esoteric. The techniques used involved such arcane terminology as "radius of search", "octant search", "shadow effect", and many others. People would argue at meetings over whether "one over the squared distance" was the appropriate decay function for individual data points. Various experiments were performed to estimate the correct exponent for such functions. Estimates were made of the optimum grid cell spacing for particular data sets, prior to contour threading through the resulting grid.

Shortly thereafter "polynomial patch" techniques came along to compete with the more traditional "weighted average" or "moving average" methods previously used, and a variety of "blending functions", "continuity constraints" and other relatives were introduced to sew the patches together and to attempt to hide the surgery.

Finally, and probably worst of all, the user of any particular contouring package was required to estimate the values for a variety of non-intuitive parameters needed for the more-or-less successful operation of the contouring programme itself. Even then problems arose– especially if the data was not as "well- distributed" as the programme developers had envisaged: rarely though had they in any way defined the assumptions under which they were working. These difficulties are discussed further in Gold (1984).

In the last few years the development of GIS and related disciplines has brought a new, and possibly better, perspective to the problem at hand, hopefully reducing many of the sources of confusion. For interpolation, as for GIS, the key question concerns the appropriate specification of spatial adjacency.

Definition of the problem

Let us first define our terms for the problem under discussion. What is required is an interpolation technique, as well as the final production (if required) of contour lines, that will have a variety of well defined properties. For the purposes of the discussion the problem is that posed by the needs of the oil industry. The issues include: a fairly sparse set of data observations; high precision and high cost for each individual observation (an oil exploration drill hole); and a highly irregular distribution of data points, as in clusters or strings. In addition, the surface must match precisely all given observations and have "reasonable" or "correct" slope values at each known location. Also the surface is to be "smooth" throughout, except where explicitly stated (Gold 1984). While this is not the only possible definition of the problem, it is sufficiently general to be appropriate for a large number of applications. In addition, the inexperienced user will probably assume properties of this kind unless informed otherwise.

Traditional methods

Let us review some of the traditional interpolation methods. Based on Newtonian ideas, the usual approach starts by throwing a grid over the map and then interpolating or estimating elevation values at each grid node. While this is not necessary– indeed it is frequently harmful in terms of producing accurate contours on the interpolated surface– it will serve for the moment as a basis for discussing the more critical point: the interpolation technique itself. Two traditional interpolation approaches have been used, each with their own advantages and disadvantages.

The first of these - the weighted average method - selects a set of neighbouring data points to the grid node under consideration and performs some averaging process on this set of observations. The averaging process includes a weighting factor that becomes larger the closer the data point is to the grid node being estimated. This approach raises some basic questions about how to select the set of neighbouring observations to be used, as well as how to select the weighting function to apply to each observation.

The second technique– using local polynomial patches– has the potential attraction that in some cases (e.g. where the patches are triangular plates between data points) the interpolated surface may readily be made to conform precisely to the slope and elevation of each data point. All polynomial patches, however, suffer from the disadvantage of a lower order of continuity ("smoothness") along patch boundaries than is present in the patch interior. Typically slopes are preserved at patch edges, but curvature is discontinuous, leading to ugly and misleading results. The inexperienced user will assume a greater level of mathematical continuity over the whole surface than is actually the case, attributing to the real data some surface features that are merely artifacts of the patch procedure.

In addition to weighted-average and polynomial-patch methods there are a variety of global techniques such as trend-surface analysis which are not usually appropriate for surface modelling where detailed examination of the surface is required. See, for example, Davis (1973). The discussion in this paper will focus on weighted-average techniques, as they appear to be the most flexible, and more readily extended to complex situations.

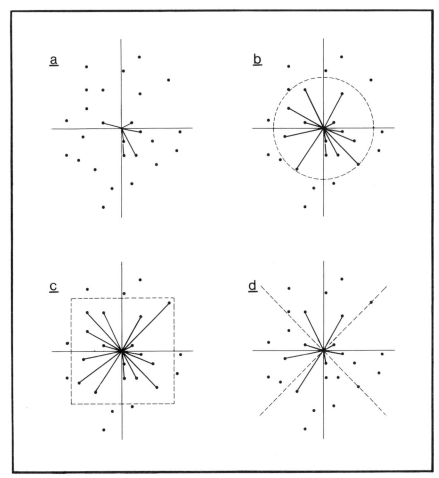

Figure 3.1 Strategies for selecting data points adjacent to a sampling location.

Spatial adjacency issues

For the weighted-average procedure, Newton's "metric" methods suggest that the set of neighbouring data points should be selected on the basis of their distance from the grid node in question. As shown in Figure 3.1 a variety of selection strategies may be used, even having been given this definition. In Figure 3.1(a) a specified number (six in this example) of closest points are collected regardless of their individual distances from the grid intersection or their distribution around that location. In Figure 3.1(b) all data points within a user-specified distance of the grid node are selected, and in Figure 3.1(c) a rectangular area is used for data point collection. These approaches might suffice if the data is isotropically distributed with a known density, but in other cases the selected points may be very poorly distributed around the grid node, as is the case in Figure 3.1(a). In Figure 3.1(d) the potential ill-distribution of selected data points is partially handled by requiring one data point to be selected in each octant around the grid node. Note also that some data points may be almost entirely "hidden" from the grid node by closer points, and for precise interpolation (as opposed to regional averaging) these should be excluded from the neighbourhood set.

In all of these cases a very important but frequently overlooked issue emerges: a minor perturbation of the grid node location or of the coordinate system definition could produce a very significant change in the set of data points selected for estimation purposes. As a result a data point with a very significant weighting, and hence a large contribution towards the final elevation estimate, could be included or excluded due to a trivial coordinate change, producing a discontinuity in the total modelled surface. A moral emerges from this: metric distance is a poor measure of spatial adjacency.

Once a set of "neighbouring" observations is selected by one of the above methods, a second problem occurs– to select the appropriate weighting function for each of these observations so that two basic conditions are obtained. Firstly, as the grid node or sampling location approaches any particular data point the surface estimation method must produce elevation estimates that converge on the true data point elevation. They must also converge in a smooth fashion, producing both elevation and slope continuity across data points. The second condition required for the weighting function is that by the time the data points are rejected from the set used for estimating the grid node value (as the grid node moves away) the weighting of the data point becomes precisely zero. Without this condition the previously-mentioned surface discontinuities will be introduced. The mathematical complexities of achieving this objective with an arbitrary selection of neighbouring data points is considerable. A second moral emerges: selection of the data points and the weighting function are interrelated processes.

The Voronoi approach to spatial adjacency

Clearly we need to be able to define a more general idea of spatial adjacency and weighting processes than a simple metric measure: we need some consistent definition of spatial adjacency. Spatial adjacency has been treated extensively in the GIS literature, primarily with respect to the linking of line segments to form polygons, but relatively little has been done to define spatial adjacency for essentially disconnected objects such as the data points used here. How are we to define spatial adjacency in this case? From current work the answer appears to be the Voronoi diagram.

What is the Voronoi diagram? Figure 3.2, from Morgan (1967), provides an illustration based on the "real world", and not just on a mathematical concept. If we put wicks in blotting

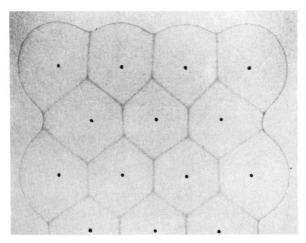

Figure 3.2 Voronoi Polygons created with blotting paper and wicks. From Morgan (1967).

paper and add solvent and ink to the wicks we will end up with a set of convex polygonal "zones of influence" as shown. The mathematical construction of the resulting point-Voronoi diagram is well described in the literature (Preparata and Shamos 1985). For our purposes the point Voronoi diagram consists of the convex polygon within which all points on the map are closer to the generating central data point than to any other data point.

Figure 3.3(a) shows the data for a well-known test set found in Davis (1973). Approximately 50 data points represent surveyed values for a small piece of terrain. Figure 3.3(b) shows the result of generating the Voronoi diagram for those points. Since all the

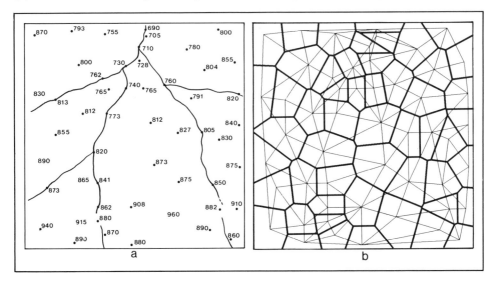

Figure 3.3 Elevation data after Davis (1973): (a) raw data, and (b) subsequent Voronoi polygons and Delaunay triangles.

boundaries between adjacent polygons are perpendicular bisectors of the lines connecting the two data points concerned, an equivalent description of the polygonal zones may be formed by the "dual" underlying triangulation, also illustrated. One possible contouring strategy would be to take each of these resulting triangular plates and pass contours through them. Figure 3.4 shows the result of performing linear interpolation in this manner. Other strategies use triangular patches to provide smooth surfaces throughout the region of interest, but the problems of these polynomial patches have already been discussed: the continuity limitations between adjacent patches have made this approach less popular than it once was.

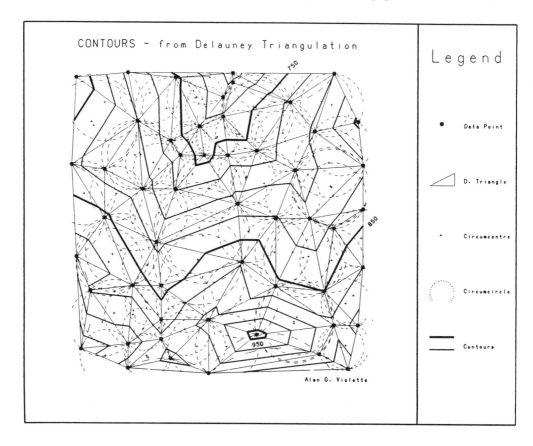

Figure 3.4 Linear interpolation of the Davis data, showing the circumcircle for each Delaunay triangle.

In practice the fundamental value of the triangulated data set (the Delaunay triangulation) is not to provide triangular "patches" but merely to provide a description of the spatial adjacency relationships between data points. Thus we may say that data points are adjacent to each other if they have a common polygonal boundary in the point-Voronoi diagram or the equivalent "dual" triangle edge.

Interpolation by theft

In order to provide weighted-average interpolation for any arbitrary grid node location we are not particularly concerned with which of the original data points are adjacent to other data points; what we want to know is– which of the data points are adjacent to the grid node under investigation? The conclusion therefore is straightforward– first you must insert your sampling location into the triangulated data set. Then you can determine the spatially adjacent data points. Figure 3.5(a) shows an initial triangulation together with the location X where a sampling data point is to be inserted. Figure 3.5(b) shows the polygon resulting from this insertion process. The data points adjacent to this sampling point are those having a common polygonal boundary with it.

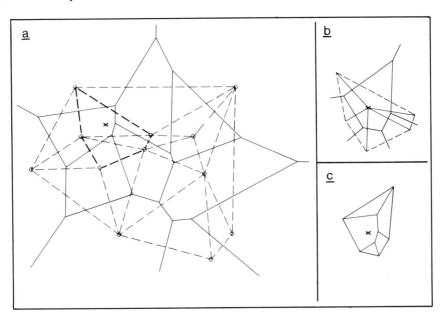

Figure 3.5   Area stealing: (a) original Voronoi polygons, plus Delaunay triangles (dashed); (b) result of inserting sampling point X; (c) areas stolen from adjacent polygons.

We thus have a definition of the adjacent data points to be used in our weighted average procedure. How do we determine the weighting functions? Our requirements for the weighting of a data point are: that it reach unity when the sampling location reaches the data point; that it reaches this in a "smooth" fashion; and that the weight of a particular data point reaches zero at the time that the data point ceases to belong to the set of adjacent points. Figure 4.5(c) shows an appropriate solution. The insertion of the sample point not only creates a new polygonal region, but it creates this region at the expense of the neighbouring data point regions. These areas stolen from the adjacent regions form the basis of our weighting function. Note that as the sampling location X approaches a data point, all of the stolen area will be obtained from that particular data point polygon, giving full weighting to that point, as required. In addition, as the sample point moves away from a particular data point the area stolen will decrease, as will the common boundary edge between the two polygons, until both of these reach zero– at precisely the time that the data point ceases to be spatially adjacent to the sample point. Thus we have achieved our main objective, and have shown that spatial adjacency and weighting are part of the same technique.

Beyond linear interpolation

Looking at this process in one independent dimension, Figure 3.6(a) shows three data points, marked D.P., along a line. Figure 3.6(b) shows their one dimensional Voronoi "zones of influence", where each building block extends to the mid-point between two adjacent data points. Figure 3.6(c) shows the introduction of our sampling point. Its zone of influence also extends to the mid-point between each of the adjacent data points, and consequently it has stolen some of the adjacent areas. Taking a weighted average of the elevations for each of the adjacent data points; multiplying by the stolen area; and finally summing them and dividing by the total area gives the estimate of elevation shown in Figure 3.6(c). This is interpolation by "conservation of area" (in two independent dimensions plus elevation, "conservation of volume").

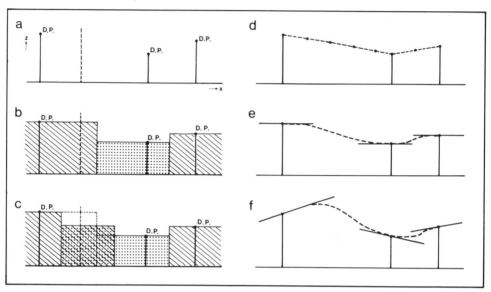

Figure 3.6 Interpolation with one independent variable: (a) original data; (b) Voronoi regions; (c) sampling point insertion; (d) resulting linear interpolation; (e) addition of blending function; (f) inclusion of slope data.

If we repeat this process for varying sample site locations, we obtain the result shown in Figure 3.6(d). We have achieved linear interpolation. Thus the first conclusion we can make is that our simple weighted average process in two dimensions is a direct analogue of linear interpolation in one dimension. There is however still one difficulty: the surface is not smoothly continuous at data points. To achieve this we need to apply a simple blending function to the individual weights before summing, as shown in Figure 3.6(e). In the current work this is a simple hermitian polynomial used to re-scale the initial linear weights. Finally, the weighted value at each data point need not be a simple elevation value. It may be any appropriate function. In Figure 3.6(f) a simple linear function is provided at each data point, and the interpolation procedure just described will guarantee a continuous surface honouring these data points and their slope functions as required. Again there is a direct analogy with interpolation techniques in one independent dimension. In passing, it should be noted that while this technique was developed independently by this writer for a commercial contouring package (Gold 1988), the use of Voronoi sub-tiles in a polynomial patch equation was first described by Sibson (1982).

The proof of the pudding– good cooks and bad

Putting this procedure into practice in two dimensions, plus an elevation dimension, Figure 3.7(a) shows a simple pyramidal data set. Figure 3.7(b) shows the result of interpolating using the procedure previously described. One point to be noted is that sub-sampling

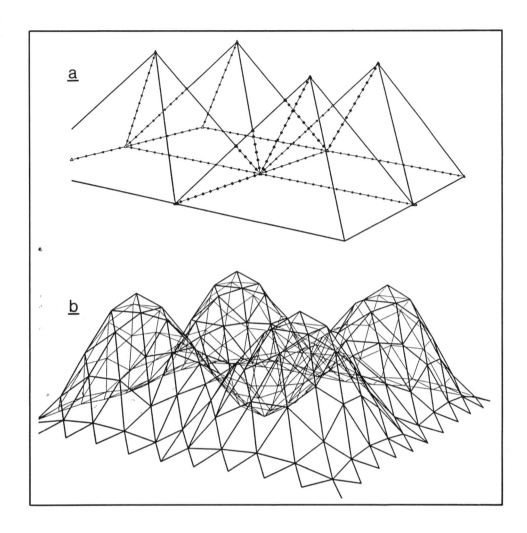

Figure 3.7 (a) Pyramid data; (b) subsequent interpolation.

locations are not based on a regular square grid, but on a regular sub-division of each of the original triangles used to form the pyramids. For details see Gold and Cormack (1987).

Not also that the surface is honoured precisely at each data point. Zero slopes are defined at each data point, and these slopes are also honoured. While the triangular subdivision used here is fairly coarse in order to illustrate the technique, an increasingly fine sub-division would illustrate that the surface preserves first order continuity or smoothness at each of the original data points.

Let us look at this process for a "real" data set. Figure 3.8 shows a particularly awkward data distribution– a set of elevations along several seismic traverses. The interpolation procedure is capable of providing a "reasonable" interpolation even in regions of the map where data points are notably absent. The surface may be considered to possess the required properties of honouring data points, preserving reasonable slopes and being appropriately smooth as well as interpolating well in sparse regions. All in all the described procedure

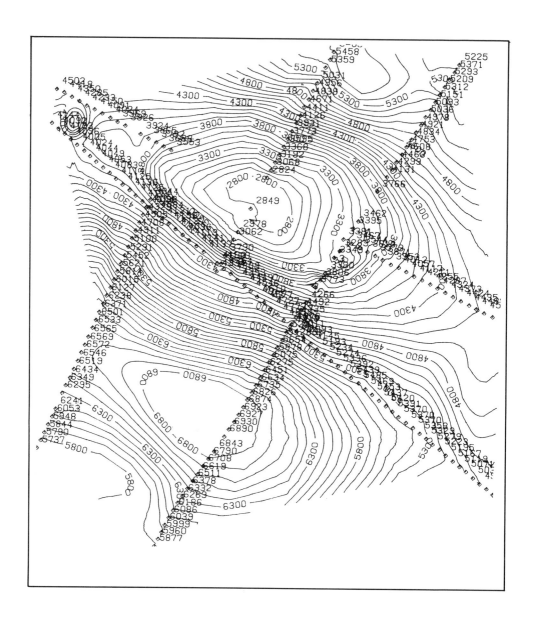

Figure 3.8 Seismic data: Voronoi interpolation.

approximates fairly well the result produced by an experienced geologist with no particular bias towards any specific subsurface interpretation.

Next the same data set is processed by two well-known commercially-available contouring packages, which shall remain anonymous. Figure 3.9(a) shows the result obtained by using a typical gridding algorithm. While the surface is recognizable, various problems, such as contour overlap, distortion in regions of low data density, and the inability to make estimates at some particular locations, are notably visible. The procedure used was a weighted average

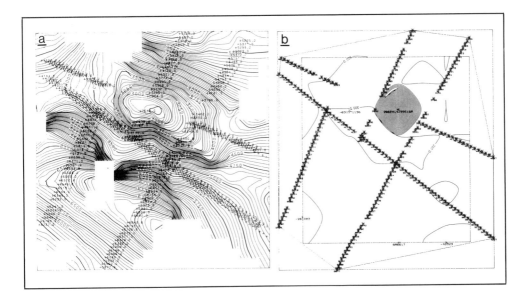

Figure 3.9 Seismic data: (a) a traditional gridding method; (b) a spline method.

of the six nearest data points. Figure 3.9(b) shows the result using another programme based on spline functions. Clearly the parameters used were inappropriate for the data set (although subsequent attempts were not necessarily much better)! While these would in no way constitute fair trials of the programmes concerned, they are good examples, if extreme ones, of the potential penalties for inappropriate specification of required system parameters. It should be noted that all three of these contour maps were produced as unsupervised student assignments, and in most cases used default system parameters. It is scarcely fair at this stage to produce another moral: metric-based systems depend on assumptions about the data, as do their default parameters. While these last maps might have been rejected outright by even an inexperienced user, less extreme errors might well have been accepted in ignorance.

Beyond points

At the conclusion of the development of this procedure we have achieved a smooth surface honouring all data points whatever their particular distribution. In some cases however smoothness is not what is required– we may have additional specific information involving breaks in slope and faults. How should these be handled?

The first consideration is the ability to handle line segments within our data set– since the features we are concerned with (breaks in slope and faults) are linear features. Figure 3.10(a) shows a portion of the Davis data set. Figure 3.10(b) shows the same portion with a linear segment added, connecting two of the previous data points. The difference between these two involves the addition of a new object, a line segment, between two existing data points.

The triangulation structure has been updated to include this. The Voronoi regions (the zones of influence) have also been modified. The fundamental definition– that all map locations within the zone of influence are closer to that object than to any other– has been maintained, but the distance-measuring function has to be modified to handle perpendicular-distance measurements from the line segment itself. The resulting zone includes linear boundaries as well as parabolic boundaries. Note that the procedure outlined here recognizes

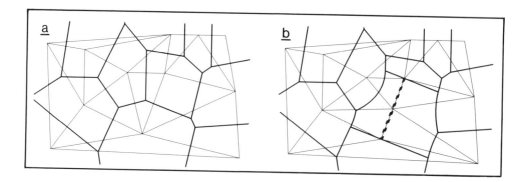

Figure 3.10 Davis data subset: (a) without, and (b) with, a linear segment (dashed line) inserted into the Voronoi diagram.

that a line (link) between two points adds information to the map and is thus itself a map object.

Beyond smoothing

Figure 3.11(a) shows a simple map with two line segment entities. Boundaries between points and points are straight lines, boundaries between points and lines are parabolas, and boundaries between lines and lines are again linear. Thus the Voronoi regions of points and line segments can be defined. In Figure 3.11(b) a sampling point has been introduced on the left hand side and the stolen areas of the adjacent polygons are shown. In Figure 3.11(c) this sampling point moves closer to one of the line segment entities and more of its stolen Voronoi region comes from the zone of the line segment object itself. In Figure 3.11(d) the sampling point is closer yet to the line segment. In all of these cases "theft" of an area can be performed for line segments as well as for point data objects.

Since area-stealing may be performed on the zones of influence of line-segment objects as well as on point objects, line-segments may be inserted as objects in the data set used to create the original map. Thus where appropriate data may be specified for a line-segment (e.g. a segment of a contour), this may be included in the data set to be subsequently used for interpolation. As with point objects, an appropriate blending function is needed if smoothness across the line-segment is desired. However, if a break in slope is required, the original linear interpolation to this object is preferable. In addition, if a fault is to be simulated, this can be done by ignoring all areas stolen from the line-segment data object. In that case interpolation would proceed independently on each side of the line segment, and the surface would become indeterminate at the line segment itself.

It is only fair to state that while the concept of area stealing for point and line segment Voronoi diagrams is easy to visualize, the implementation of this point and line Voronoi diagram is extremely difficult and abounds in special cases. Nevertheless, the techniques discussed here provide a consistent class of interpolating functions appropriate for a wide variety of general applications.

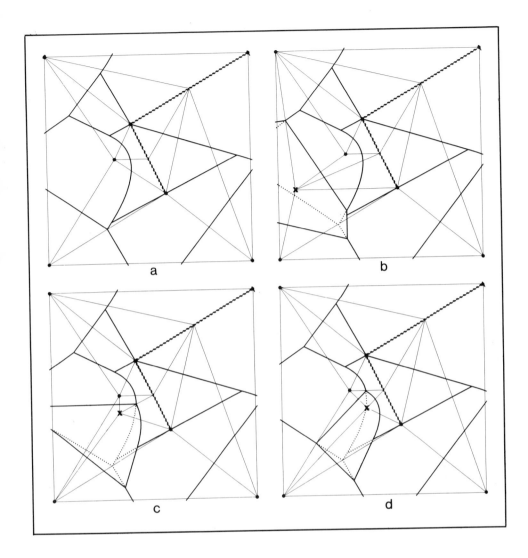

Figure 3.11 Point and line Voronoi diagrams: (a) shows seven points (dots) plus two line segments (dashed), together with the dual Delaunay triangulation; (b), (c) and (d) show insertion of a sampling point X and the resulting stolen areas (dotted lines show regions stolen from the original data set in (a)).

Conclusions

To return to the original Davis data set, Figure 3.12(a) illustrates a perspective view of this data set looking from the north.

Initially only a simple linear interpolation over each triangular plate is implemented. In Figure 3.12(b) subdivision of the original triangular facets is performed and interpolation (as previously described) performed at each of these sub-nodes. While the surface represented is accurate, in the sense of honouring data point information, it is difficult to perceive the surface form due to a lack of perspective depth cues. In Figure 3.12(c) the surface of Figure 3.12(b) has been re-sampled on a regular grid and re-displayed, giving a simple perspective

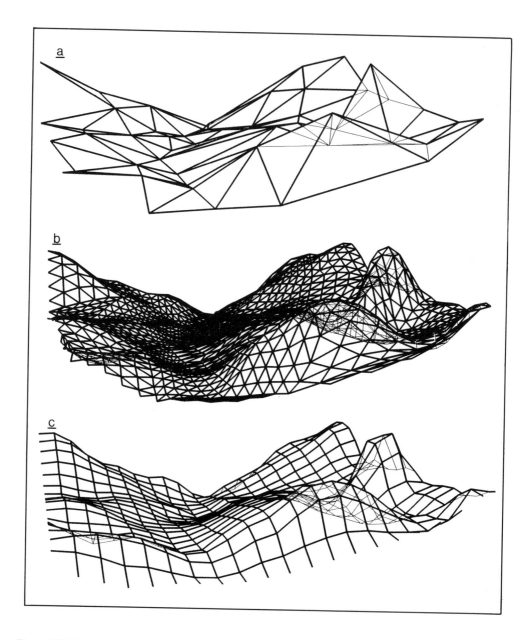

Figure 3.12. Davis data: (a) simple triangulation; (b) Voronoi interpolation on sub-triangles; (c) Voronoi interpolation on a grid.

view. Note, however, that there is no longer any guarantee that the original data points are, or should be, honoured in detail.

In conclusion I would like to emphasize two points. The first and more specific one is that the interpolation problem can properly be phrased as a spatial adjacency problem. The techniques described here appear to open up a great variety of applications not available with other methods. The second point to be made is that the trend both in this work and in GIS. in general is towards better and better understanding of the spatial relationships between objects, as well as their coordinate locations. The methodology described here is not restricted to two spatial dimensions, although the introduction of line segments into the three-dimensional point Voronoi diagram has not yet been attempted. Nevertheless, the Voronoi concept may well provide a consistent basis for understanding the spatial relationships of various types of map objects in a varying number of dimensions.

Acknowledgements

The funding for this research was provided in part by an operating grant from the Natural Sciences and Engineering Research Council of Canada, and in part from the Energy, Mines and Resources Canada Research Agreement Programme.

References

Davis, J. C., 1973, *Statistics and data analysis in geology*, (New York: John Wiley and Sons), 313p.

Gold, C.M., 1984, Common-sense automated contouring– some generalizations. *Cartographica*, **21**, 121–129.

Gold, C. M., 1988, Point and area interpolation and the digital terrain model. In *Proceedings of the Second International Seminar on Trends and Concerns of Spatial Sciences*, (Fredericton: University of New Brunswick), pp. 133–147.

Gold, C. M. and Cormack, S., 1987, Spatially ordered networks and topographic reconstruction. *International Journal of Geographical Information Systems*, **1**, 137–148.

Morgan, M. A., 1967, Hardware models in geography. In *Models in Geography*, Chorley, R. J. and Haggett, P., pp. 727-774, (London: Methuen).

Preparata, F. P. and Shamos, M. I., 1985, *Computational Geometry*, (New York: Springer-Verlag), 390p.

Sibson, R., 1982, A brief description of natural neighbour interpolation. In *Interpreting Multivariate Data*, Barnett, V., pp. 21-36. (London: John Wiley and Sons)

Chapter 4

Efficient mapping of heavy metal pollution on floodplains by co-kriging from elevation data

Henk Leenaers, Peter A. Burrough and Joop P. Okx

Introduction

Regionalized variable theory (Matheron, 1971) has been much used recently for making quantitative maps of single or multiple soil properties (Burgess & Webster, 1980; Burrough, 1986; Davis, 1986; Flatman & Yfantis, 1984; Gilbert & Simpson, 1985; Journel & Huijbregts, 1978; Laslett et al. 1987; McBratney & Webster, 1986; Oliver & Webster, 1986; Starks et al, 1987; Webster, 1985). The theory provides a convenient means of summarizing soil spatial variability in the form of an auto (or single property) variogram that can be used to estimate weights for interpolating the value of a given soil property at an unsampled location. Kriging (as this technique is known) is a form of weighted local averaging that is optimal in the sense that it provides estimates of values at unrecorded sites without bias and with minimum known variance. Co-kriging is the logical extension of kriging to situations where two or more variables are spatially interdependent and the one of immediate interest is undersampled (David, 1977; Journel & Huijbregts, 1978; McBratney & Webster, 1983; Vauclin et al, 1983). Co-kriging could be useful for mapping a property that is expensive to measure by making use of a large spatial correlation between the property of interest and some cheaper-to-measure attribute.

Because of past mining activities, the floodplain soils of the Geul river are polluted with heavy metals, particularly zinc, lead and cadmium (Leenaers et al, 1988; Rang et al, 1986). The general pollution pattern consists of a logarithmic decay with distance to the source of contaminants (Leenaers et al, 1988), with local deviations that are caused by variations of flood frequency and by differences in sedimentary conditions during flood events. The pollutants constrain the land use in these areas, so detailed maps are required that delineate zones with high concentration levels. Successful attempts have been made by Wolfenden & Lewin (1977) and Rang et al. (1987) to relate the pollution level of floodplain soils to floodplain characteristics such as fluviatile landform, inundation frequency and soil type. These studies led to the production of choropleth maps that delineated broad zones for which the measures of central tendency and variation of a certain pollutant were known. Detailed information about the continuous spatial variation of pollution levels within the zones, however, was not provided by these studies.

Detailed planning studies in South Limburg required that the distribution of heavy metal pollution be mapped both quantitatively and as accurately as possible but collecting data on the metal content of soil material is laborious and expensive. Therefore mapping the heavy metal contents directly would be extremely expensive. The observation of a close correspondence between the relative elevation of the floodplain and heavy metal pollution levels suggested that elevation data could be used to advantage to support the mapping. The data on relative elevation can be gathered cheaply and quickly either in the field or from spot heights on 1:10 000 maps (Netherlands Topographic Survey, scale 1:10.000, 1976).

The aim of this study was to investigate the improvements in accuracy and cost-effectiveness resulting from using co-kriging with elevation data to map heavy metal pollutants compared with simple kriging or with simple linear regression based on the correlation between metal content and relative elevation. The three techniques were used to map of zinc concentrations in the floodplain soils from laboratory measurements of zinc concentrations sampled at a limited number of sites. These data were supplemented by elevation data for the co-kriging and regression mapping. The maps were tested with an independent subset of the original data (45 out of 199 sample sites).

The study area

The Geul is a tributary of the Meuse. From its source to its confluence with the Meuse it is 56 km long, of which 36 are in Belgium, and it drops 242 m. The catchment covers 350 km (Figure 4.1). The discharge of the Geul largely depends on rainfall. At the Dutch-Belgian border the average flow is 1 m³/s and the maximum discharge is about 30 m³/s. These values increase to 3 and 60 m³/s, respectively, near the confluence with the Meuse.

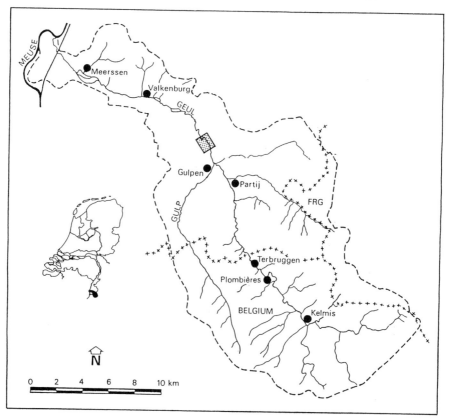

Figure 4.1 Location of the study area.

In the southern part of the Netherlands the Geul valley incises a loess-covered limestone (chalk) plateau. Cultivation of the forested valley slopes began in Roman times. Ever since soil erosion has supplied loess materials, which are rich in the silt and fine sand fraction, to the Geul resulting in floodplain deposits with a relatively coarse texture containing only 10-30 % clay.

Important occurrences of heavy metal ores are found near Plombieres and Kelmis, both situated in the Belgian part of the Geul basin. Exploitation of the zinc and lead ores probably began in the 13th century. The heyday of mining was 1820-1880 and the last mine closed in 1938. Separation techniques, exploiting differences in specific density were used when mining the ores. These techniques were inefficient and resulted in high concentrations of ore particles and metal-rich spoil in the effluent, which was discharged directly into the river. The reject material and tailings were dumped in large heaps along the riverbanks. Some of these heaps still exist. An additional source of metals is formed by erosion of older, locally highly contaminated streambank deposits.

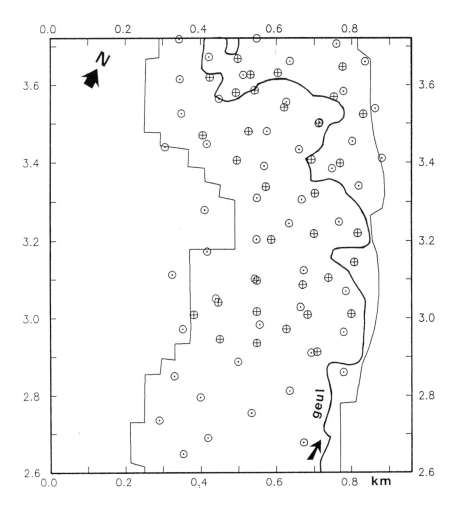

Figure 4.2 Sample locations in part of the study area (⊕: known Zn and RE; O: known RE).

Experimental and analytical procedures

Field methods

Samples of topsoil (0-10 cm; 100 g) were collected at 199 sample sites in a 5 km long part of the floodplain area of the Geul near the Dutch town of Gulpen where the valley is 300-600 m wide. The distances between sample locations range from 50-100 m perpendicular to the valley axis to 200-600 m parallel to the valley axis. The sample locations in the part of the study area are shown in Figure 4.2. Because of the shape of the study area— its length is ten times as large as its width— it is impractical to show all data. This study presents results for the part of the study area shown in Figure 4.2.

Laboratory methods

All 199 sediment samples were dried for 24 hours at 60°C, and crushed in a mortar. One gram of this material was boiled gently with 20 ml 30 % HNO3 for two hours. The extract was then separated from the sediment by centrifuging and brought to 40 ml with distilled water. The concentrations of lead, zinc cadmium and copper were high enough to be determined by direct flame absorption spectometry. In this paper only the spatial distribution of zinc will be studied.

Elevation data and sample sites

The Netherlands Topographic Survey provides maps at scale 1:10000 that show not only the contour lines but also accurately measured spot heights. Within the study area, the elevation was known directly for 309 spot heights, which did not coincide with the soil sample sites. These spot heights were supplemented by data from digitized contours to yield a total elevation data set for the study area of c. 3000 points. These data include the trend in elevation of the long profile of the river. The relative elevation (RE) of the 199 soil sample sites relative to the river bed was determined from the elevation data set by removing the trend due to the river profile and by local linear interpolation. This preparation gave a data set for the 199 sample sites where both zinc level and relative elevation were known. For co-kriging, the RE values at the 199 sample points were supplemented by the relative elevations computed by subtracting the trend from the absolute elevation at the 309 spot heights. Together with the heights at the soil sample sites they formed a set of 508 points at which the RE was known.

The set of 199 data points at which zinc levels and elevation were known was split into two. The larger set contained 154 data points that were used for computing the variograms for zinc and for the cross variogram between zinc and RE and for the point kriging of zinc. The smaller data set contained 45 data points that were to be used for validating the interpolations. For co-kriging, the 154 data points were supplemented by the 309 spot heights; the variogram of RE was also computed from these 463 RE data. The reason these extra RE data were used was to make as much use as possible of the cheap, readily available data.

Preliminary data analysis

Table 4.1 gives parameters of the frequency distributions of relative elevations (n=463) and of zinc concentrations (n=154) in the alluvial deposits. The degree of pollution of the floodplain soils can be seen from the parameters of a set of samples (n=12) taken from non-flooded colluvial loess sediments in the Geul valley, which are also in the table.

	n	mean	median	minimum	maximum	variance
Zinc content of alluvial sediments	154	741	543	114	2270	292877
Zinc content of colluvial sediments	12	147	124	68	338	5112
Relative elevation	463	429	411	224	791	9032

Table 4.1 Measures of central tendency and variation of relative elevations (cm) and of zinc concentrations (mg/kg) in alluvial and colluvial sediments in the Geul valley.

The zinc concentrations in the alluvial deposits are clearly much higher than the concentrations in sediments that are not influenced by the presence of a river. Moreover, variability of the zinc concentrations in the flood deposits is much greater than in the colluvium because of differences in the frequency of flooding and in sedimentary conditions during flood events. Regression analysis of zinc concentration versus distance downstream revealed that along this short distance of 5 km the decay of metal concentrations in soil material with distance parallel to the river is not significant.

Table 4.2 lists the correlation coefficients of Zinc (Zn) and the $\log_{10}$ of Zinc with relative elevation (RE) and the $\log_{10}$ of relative elevation. The correlations are statistically significant but lower than preliminary field studies had suggested.

Y	X	a	b	r
Zinc	RE	1617	-206	-0.38
$\log_{10}(Zn)$	RE	3.48	-0.17	-0.48
Zn	$\log_{10}(RE)$	5986	-2004	-0.36
$\log_{10}(Zn)$	$\log_{10}(RE)$	7.19	-1.70	-0.46

Table 4.2 Correlation coefficients and parameters of linear relations (Y=a+bX) between zinc concentrations (Zn) and relative elevation (RE) (n=154).

Methods used for mapping

The variation of zinc concentration over the test area was investigated using three methods:
(a) Linear regression from the 45 RE data at the test locations (using the strongest relation in Table 4.2);
(b) By point kriging from the reduced set of Zn data (n=154); and
(c) By point co-kriging from the reduced set of zinc data (n=154) and the reduced set of data on relative elevation (n=463).

Linear regression

Procedure (a) was not used to make a map; the results were evaluated by computing the absolute squared estimation errors between estimated and measured values of zinc at the 45 data points (see Table 4.4).

Point kriging

Procedure (b) was used to make a map. The procedure of point kriging is well-known (see Burgess & Webster 1980) and will not be further described here. Variograms for both zinc content and RE were computed using standard procedures (Figures 4.3(a,b)). Each calculated semivariance is plotted with a symbol that represents a distance class of 0.7 km. An exponential model provided a good fit to all experimental variograms. The equation is:

$$\gamma(h) = C_0 + C(1 - e^{(-h/a)}) \text{ for } h > 0 \tag{1}$$

$$\gamma(0) = 0 \tag{2}$$

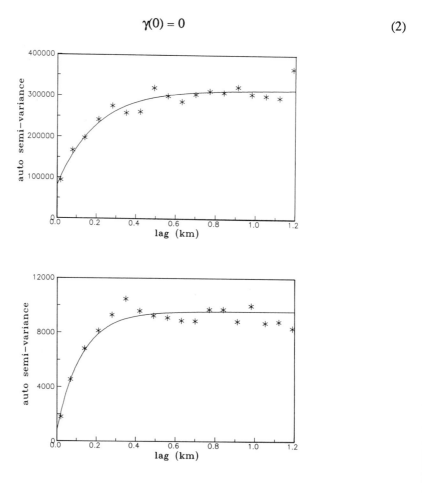

Figure 4.3(a,b) Auto variograms of top soil zinc content and relative elevation.

where a is a constant, C_0 is the nugget variance and $C_0 + C$ is the sill. The practical range a' may be defined as (Journel & Huijbregts, 1978):

$$a'=3a \text{ for which } \gamma(a') = C_0 + C(1-e^{-3}) = C_0 + C(0.95) \approx C_0 + C \qquad (3)$$

The parameters of the variogram models were fitted by a least squares approximations (Cressie, 1985). The results are listed in Table 4.3. They show that the topsoil zinc content is highly variable and has a nugget variance that is approximately 25 % of the sill. The relative elevation is much less variable and has a smaller nugget variance of approximately 9 % of the sill.

attribute(s)	nugget(C_0)	sill(C_0+C)	range(a')
Zn	379262	314958	0.59
RE	849	9564	0.38
Zn–RE	-4254	-20351	0.73

Table 4.3 Parameters of the theoretical variogram functions.

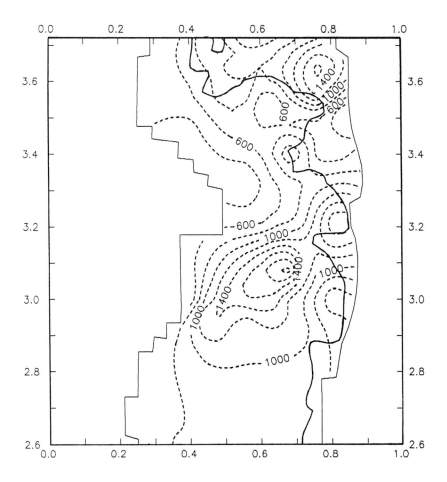

Figure 4.4 Interpolated map of top soil zinc content by point kriging.

The lags at which both variograms level out to reach the sill are in the same order of magnitude, i.e. 0.4-0.6 km. Obviously, the sill variance of each property has the same order of magnitude as the total variance (see Table 4.1).

The variogram model for zinc was used to interpolate the values of zinc at the nodes of a 40 x 40m grid for an area measuring 1.4 x 5.0 km (Figure 4.4). The map was tested using the 45 test sites as above (see Table 4.4).

Co-kriging

Procedure (c) was also used to make a map. Because the co-kriging method is less well documented than point kriging, we give a description of the methods used for estimating the cross variogram and for interpolation.

The cross variogram

Just as values of a property can depend in the statistical sense on those of the same property at other places nearby, so can they be related spatially to values of other properties. Where this is so the variables are said to be co-regionalized: they are spatially dependent on one another. By analogy with the single variable, the dependence between two variables can be expressed by a cross variogram. For any pair of variables U and V, the cross semi-variance $\gamma_{UV}(h)$ at lag h is defined as:

$$2\,\gamma_{UV}(h)=E[\{z_U(x)-z_U(x+h)\}\{z_V(x)-z_V(x+h)\}] \tag{4}$$

where z_U and z_v are the values of U and V at places x and x+h. If U=V the above equation denotes the auto variogram (McBratney & Webster, 1983).

The cross variogram is estimated directly from the sample data using the equation:

$$\gamma_{UV}(h)= \frac{1}{2N(h)} \sum_{i=1}^{N(h)}\{z_u(x_i)-z_u(x_i+h)\}\{z_v(x_i)-z_v(x_i+h)\} \tag{5}$$

where N is the number of data pairs at locations x_i and x_i+h in a given distance and direction class h (David, 1977).

In the process of cross variogram fitting the Cauchy-Schwarz inequality:

$$|\gamma_{UV}(h)| \leq \sqrt{\gamma_U(h)^* \, \gamma_V(h)} \text{ for all } h>=0 \tag{6}$$

was checked so as to guarantee a positive cokriging variance in all circumstances (Journel and Huybregts, 1978; Myers, 1982; Myers, 1984; Nienhuis, 1987). Figure 4.5 shows the cross variogram. The topsoil zinc content is negatively correlated with the relative elevation (see Table 4.2), so the cross variogram is negative. The nugget variance is approximately 21 % of the sill.

Anisotropy effects may occur if the spatial dependency of any variable has characteristics that depend on the direction for which the semi-variances are computed. In this case study, however, because the area is 5 km long but only 0.6 km wide, only one important direction (parallel to the valley axis) can be recognized. Therefore anisotropy effects were not investigated.

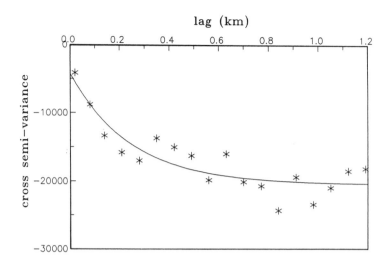

Figure 4.5 Cross variogram of relative elevation and top soil zinc content.

The co-kriging equation

A co-kriged estimate is a weighted average of the available data with weights chosen so that the estimate is unbiased and has minimum variance, and in practice only near observations carry enough weight to have effect (McBratney & Webster, 1983). If there are V variables, $v=1,2,....,V$ and each is measured at n_v places, x_{iv}, $i=1,2,....,n_v$, then the value of one of the variables, say u at x_0 is estimated by:

$$\hat{z}_u(x_0) = \sum_{v=1}^{V} \sum_{i=1}^{n_v} \gamma_{iv}\, z(x_{iv}) \text{ for all v.} \tag{7}$$

(McBratney & Webster, 1983). To avoid bias, i.e. to ensure that $E[z_u(x_0)-\hat{z}_u(x_0)]=0$, the weights, γ_{iv}, must sum as follows:

$$\sum_{i=1}^{n_v} \gamma_{iv} = 1 \text{ for v=u and } \sum_{i=1}^{n_v} \gamma_{iv} = 0 \text{ for all } v \neq u . \tag{8}$$

The first condition implies that there must be at least one observation of the variable u for co-kriging to be possible. Subject to these conditions the weights are chosen to minimize the variance,

$$\sigma_u^2(x_0) = E[\{z_u(x_0)-\hat{z}_u(x_0)\}^2] \tag{9}$$

by solving the appropriate kriging equations. There is one such equation for each combination of sampling site and property. Therefore for estimating variable 1 at site x_0 the equation for the g-th observation site of the v-th variable is:

$$\sum_{l=1}^{V} \sum_{i=1}^{n_v} \lambda_{il} \, \gamma_{il} \, (x_{il}, x_{gv}) + \varphi_v = \gamma_{uv} \, (x_0, x_{gv}) \qquad (10)$$

(McBratney & Webster, 1983) for all g=1 to n_v and all v=1 to V, where φ_v is a Lagrange multiplier. Together these equations form the co-kriging system. In this case of point co-kriging the place to be estimated is a volume of soil with the same size and shape as those on which the original observations were made.

The zinc levels were estimated for the same grid that had been used for point kriging (Figure 4.6). The map was tested in the same way (Table 4.4).

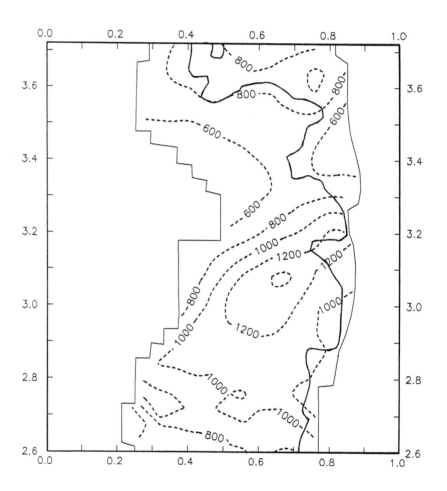

Figure 4.6 Interpolated map of top soil zinc content by co-kriging.

Comparing the 3 methods

Table 4.4 gives the results of comparing the estimates obtained by the three methods with the known zinc concentrations at the 45 test sites. The table reveals notable differences between the performance of the three methods. Both kriging methods outperform linear regression.

Furthermore, despite the weak correlation between topsoil zinc and relative elevation, the co-kriging procedure has managed to take advantage of that relation giving smaller estimation errors than simple kriging.

		Median	Interquartile range
linear regression	(AE)	353	453
	(SE)	124,357	351,102
kriging	(AE)	311	400
	(SE)	97,014	243,853
co-kriging	(AE)	283	410
	(SE)	80,241	234,954

AE: absolute errors; SE: squared errors

Table 4.4 Median and interquartile range (mg/kg) of absolute and squared prediction errors (n=45).

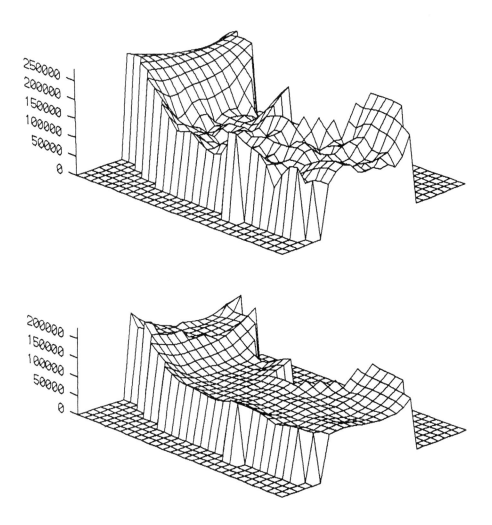

Figure 4.7: Estimation variance of Zn: a) point kriging; b) point co-kriging (view from the north).

The maps of the study area, as interpolated by kriging and cokriging (Figures 4.4 and 4.6), have a similar pattern of contour lines, but the map produced by simple kriging accentuates the extreme values (e.g in the vicinity of (0.8, 3.6) and (0.7, 3.1)) in terms of both the absolute value of the peak and the size of the area that it covers. It is difficult to judge this discrepancy. The detection of extreme values certainly is a virtue, but the overestimation of the size of its area (probably due to the lack of data on RE) is less desirable.

The value of co-kriging over simple kriging is evident in the southern part of the area where few data on zinc content were available. Because of the scarcity of data in the immediate neighbourhood, the kriging estimates of zinc form a smooth surface that tilts towards the mean zinc content of the data set (e.g. 741 mg/kg; see Table 4.1). In contrast the co-kriged map shows more detail and the estimates are larger than the mean. These estimates have been improved as a result of the presence of more data on the co-variable, the relative elevation. However, because the contour interval and the number of lines that is displayed are chosen arbitrarily, the maps need to be interpreted with caution.

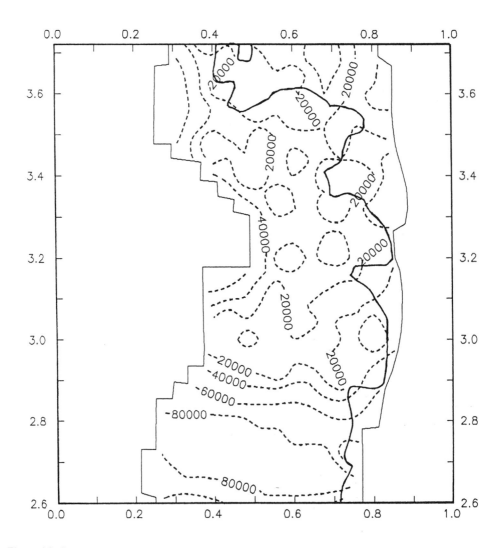

Figure 4.8 Contour map of the difference of variance between Fig 4.4a and Fig 4.4b.

The differences between the results obtained by simple kriging and by co-kriging can be seen in Figures 4.7 and 4.8, which show respectively block diagrams of the estimation variances of both methods and contour lines of the difference between the two block diagrams. The figures show the reduction of the estimation variance achieved by the co-kriging procedure. In the northern part of the area the estimation variance is quite stable and is about 10-20 % less than the variance of the kriged estimates. For both methods it can be observed that in the southern part of the area the estimation variance is considerably larger. There, the discrepancy between the methods is even more pronounced: the estimation variance obtained by co-kriging being only 70 % of that obtained by kriging.

Conclusion

Elevation data can be used to advantage in a co-kriging procedure for mapping alluvial topsoils polluted with zinc in the floodplain of the Geul. Because elevation data are readily available on the altitude map of the Netherlands Topographic Survey the improvements cost very little additional expense. Despite the weak correlation ($r=-0.37$) between topsoil zinc and relative elevation it was possible to obtain more accurate estimates of zinc content by point co-kriging than by point kriging or by simple linear regression. Moreover, the variances of the estimates of point co-kriging are substantially less than those obtained by point kriging.

Acknowledgements

This study was financed by the Netherlands Organization for Scientific Research (N.W.O.). We thank the following persons for their assistance: A. Stein (Agricultural University of Wageningen, the Netherlands), P. Nienhuis (Free University of Amsterdam), and J. van Keulen for software and geostatistical advice; P.L. Karssemeijer for the digitizing, M. Vranken who processed the elevation data, and A. Bloem, B.R. Doeve and J.W.J. van Zeijl who did the field work and the laboratory analysis. This article is a revised version of a paper first presented at the Third International Geostatistics Congress at Avignon, France, September 5-9, 1988, and is reproduced here with permission from Kluwer Publishers (The Netherlands).

References

Burgess, T. M. and Webster, R., 1980, Optimal interpolation and isarithmic mapping of soil properties: I. The semi-variogram and punctual kriging. *Journal of Soil Science*, **31**, 315–331.

Burrough, P. A., 1986, *Principles of Geographical Information Systems for Land Resources Assessment.* (Monographs on soils and resources survey 12). (Oxford: Clarendon Press), 193p.

Cressie, N., 1985, Fitting variogram models by weighted least squares, *Mathematical Geology*, **17**(5) 563–586.

David, M., 1977, *Geostatistical Ore Reserve Estimation.* Amsterdam: Elsevier, 364p.

Davis, J. C., 1986, *Statistics and Data Analysis in Geology.* (New York: John Wiley & Sons), 646p.

Dubrule, O., 1983, Cross validation of kriging in a unique neighborhood, *Mathematical Geology*, **15**(6), 687–699.

Flatman, G. T., and Yfantis, A. A., 1984, Geostatistical strategy for soil sampling: the survey and the census, *Environmental Monitoring and Assessment*, **4**, 335–349.

Gilbert, R. O., and Simpson, J. C. 1985, Kriging for estimating spatial patterns of contaminants: potential and problems, *Environmental Monitoring and Assessment*, **5**, 113–135.

Journel, A. G. and Huijbregts, A., 1978, *Mining Geostatistics*. (New York: Academic Press), 600p.

Laslett, G. M., McBratney, A. B., Pahl, P. J. and Hutchinson, M. F., 1987, Comparison of several spatial prediction methods for soil pH, *Journal of Soil Science*, **38**, 325–341.

Leenaers, H., Rang, M. C. and Schouten, C. J., 1988, Variability of the metal content of flood deposits, *Environmental Geology and Water Science*, **11**(1), 95–106.

Matheron, G., 1971, The theory of regionalized variables. *Les Cahiers du Centre de Morphologie Mathematique de Fontainbleu*, **5**, 1-210. Ecole Nationale Superieure des Mines de Paris.

McBratney, A. B. and Webster, R., 1983, Optimal Interpolation and isarithmic mapping of soil properties: V. Co-regionalization and multi-sampling strategy, *Journal of Soil Science*, **34**, 137–162.

McBratney, A. B. and Webster, R., 1986, Choosing functions for semivariograms of soil properties and fitting them to sampling estimates, *Journal of Soil Science*, **37**, 617–639.

Myers, D. E., 1982, Matrix formulation of cokriging, *Mathematical Geology*, **14**(3), 249–257.

Myers, D. E., 1984, Cokriging: new developments, In *Geostatistics for Natural Resources Characterization (part 1)*. NATO ASI Series, Series C: Mathematical and Physical Sciences, 122, Verly, G., David, M., Journel, A. G. and Marechel, A. (eds.)pp. 295–305. (Dordrecht: D. Reidel Publishing Company)

Netherlands Topographic Survey, 1976, Altitude map of the Netherlands (scale 1:10.000, **62**).

Nienhuis, P. R., 1987, CROSSV, a simple FORTRAN 77 program for calculating 2-dimensional experimental cross-variograms, *Computers & Geosciences*, **13**(4), 375–387.

Oliver, M. A. and Webster, R., 1986, Semi-variograms for modelling the spatial pattern of landform and soil properties. *Earth Surface Processes and Landforms*, **11**, 491–504.

Rang, M. C., Kleijn, C. E. and Schouten, C. J. 1986, Historical changes in the enrichment of fluvial deposits with heavy metals, *International Association of Hydrological Sciences Publication*, **157**, 47–59.

Rang, M. C., Kleijn, C. E. and Schouten, C. J. 1987, Mapping of soil pollution by application of classical geomorphological and pedological field techniques. In *International Geomorphology*, part I, Gardiner, V. (ed.), 1029–1044. (New York: Wiley & Sons).

Starks, T. H., Sparks, A. L. and Brown, K. W., 1987, Geostatistical analysis of Palmerton soil survey data, *Environmental Monitoring and Assessment*, **9**, 239–261.

Vauclin, M., Vieira, S. R., Vachaud, G. and Nielsen, D. R., 1983, The use of co-kriging with limited field soil observations, *Soil Science Society America Journal*, **47**(2), 175-184.

Webster, R., 1985, *Quantitative Spatial Analysis of Soil in the Field*. (New York: Springer-Verlag), 69p.

Wolfenden, P. J. and Lewin, J., 1977, Distributions of metal pollutants in floodplain sediments. *Catena*, **4**, 309–317.

Chapter 5

The application of a digital relief model to landform analysis in geomorphology

Richard Dikau

Introduction

In the last few years the application of GIS technologies has provided geomorphographical research with a series of new possibilities for quantitative relief form analysis. Particular emphasis has been put on the geomorphometrical point attribute approaches (Evans 1980) and the extraction of drainage basin variables from digital elevation models (DEM) (Mark 1984, Band 1986, Jenson & Domingue 1988). The automated derivation of landforms has become a necessity for quantitative analysis in geomorphology (Douglas 1986, Pike 1988). Furthermore, the application of GIS technologies has become an important tool for data management and numerical data analysis for purposes of geomorphological mapping. From the viewpoint of a geomorphological mapping project this includes the:
(1) storage of digitized spatial data from different layers of map information;
(2) evaluation of these data for applied purposes and for other geoscientific disciplines;
(3) simplification of the mapping process by modelling landform surfaces; and
(4) modelling of geomorphodynamical processes on the basis of quantified objects.

These four points summmarize the contents of a research project which aims at the development of a digital geomorphological base map (DGmBK) which can serve as a geomorphological information system. Its central function consists in the computer-aided classification and modelling of landforms on the basis of DEMs. The present paper aims to present some of the results obtained so far. Systematically the research project is based on the normalization of the Geomorphological Map of the Federal Republic of Germany 1:25000 (GMK 25) (Barsch & Liedke 1980, 1985).

Accordingly the modelling has to take into account the systematical requirements and the practical experiences of the GMK 25 project. The modelling results can thus be checked empirically in the field and with the help of the maps, of which 27 sheets from different type regions currently exist. The concept is also being tested in geoecological and pedological applications.

At first the systematical basis and some assumptions for its conversion into the computer-aided model are discussed. Then, case studies will exemplify our theories.

Hierarchical typology of geomorphographical relief units

In the context of the general discussion about 3-dimensional GIS, our current work aims to model a 2-dimensional land surface, or, in the German use of terminology, the relief or the georelief, within 3-dimensional space. In this sense the relief is regarded as the surface of a 3-dimensional object consisting of subsurface material. The relief is a highly complex continuum. A classification of this surface has to take into consideration different aspects relevant to its recent functionality and historical genesis. In terms of geomorphological research these aspects comprise the investigation of geomorphodynamic processes and the form-process reconstruction.

An important tool to normalize the spatial and genetical diversity of geomorphological objects has been developed in analytical taxonomies, as for example, by A. Penck (1894), Savigear (1965), Barsch (1969), Kugler (1974) and Speight (1984). They have been applied especially in geomorphological mapping systems due to the fact that the map and legend are a result of, and a tool for, geomorphological analysis. For example the system used by GMK 25 classifies the georelief as consisting of components ('building stones') which are characterized mainly by geomorphographics, geomorphostructure, geomorphodynamics, and geomorphogenetics. For the definition of the semantic data model of the DGmBK these components are regarded as basic object types. This current research is concerned with the geometrical and typological objects of the geomorphographical section.

A specific aspect of landform classification is shown in Figure 5.1 (Ahnert 1988), where approximate age is plotted against size of geomorphological objects. The situation here relates to the two hierarchical arrangements of time and space. The hierarchy of space can be defined by different levels or types of relief, ranging from the pico- to the mega- relief. Landforms of different sizes, from impact craters of raindrops to extensive shields, are associated with defined levels. The primary hierarchical level of our experiments comprises the micro- and meso- relief. The spatial scale of this types ranges from approximatly some meters to one kilometer. A more detailed explanation of the different size types of objects in relation to width, area and height is shown in Table 5.1.

	Main type of size order			Type of size order			
	W(m)	A(m²)	H/D(m)	W(m)	A(m²)	H/D(m)	
MEGARELIEF	$>10^6$	$>10^{12}$		$>10^6$	$>10^{12}$		Canadien shield
	— 10^6	— 10^{12}		— 10^6	— 10^{12}		
B MACRORELIEF A			$>10^3$	— 10^5	— 10^{10}	$>10^3$	Mountain area, the Alps, the Rhine Graben
	— 10^4	— 10^8	— 10^3	— 10^4	— 10^8	— 10^3	
B MESORELIEF A				— 10^3	— 10^6	— 10^2	Valley, moraine, hills, cuesta scarps
	— 10^2	10^4	10^1	— 10^2	— 10^4	— 10^1	
B MICRORELIEF A				— 10^1	— 10^2	— 10^0	Gully, ice-wedge, doline, dune, terrace
	— 10^0	— 10^0	— 10^{-1}	— 10^0	— 10^0	— 10^{-1}	
NANORELIEF							Karren, tafoni, erosion rills
	— 10^{-2}	— 10^{-4}	$<10^{-1}$	— 10^{-2}	— 10^{-4}	$<10^{-1}$	
PICORELIEF	$<10^{-2}$	$<10^{-4}$		$<10^{-2}$	$<10^{-4}$		Glacial striations

W = width of unit   A = area of unit   H/D = height/depth of unit

Table 5.1 Size order types of relief units.

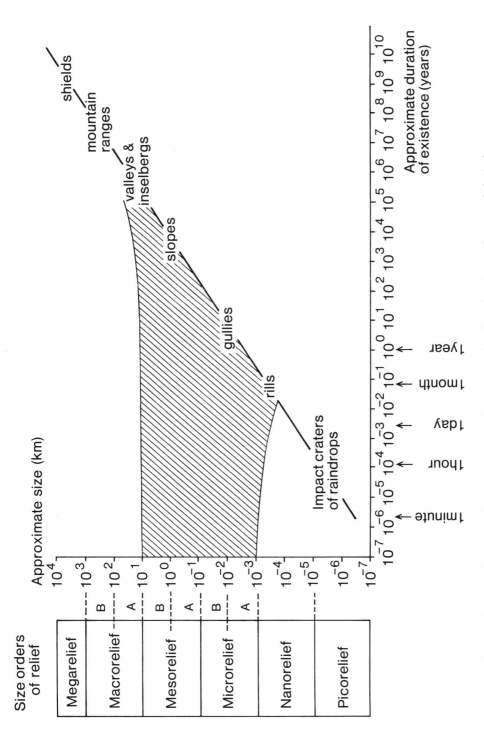

Figure 5.1 Estimation of the relationship between size types and duration of existance of geomorphological objects (after Ahnert 1988, modified).

One method to classify the high variety of relief forms discussed above is based on their hierarchical subdivision into relief units which can be described with geometrical and topological attributes. The structure of this approach is visualized in Figure 5.2 on the mesoform level. The model kernel is defined by mesoform elements and facets. These are the units for form synthesis. Their typology is strictly based on the concept of homogeneity of geometrical attributes. They can be described as surface units in 3-dimensional space, making a complete description of the surface possible. Some definitions:

Form facets are relief units with homogeneous gradient, aspect and curvature. They represent the lowest hierarchical level of every size type.

Form elements are relief units of homogeneous plan and profile curvature. They can be subdivided into form facets.

Both units may be superimposed by microforms or microform associations. Further possibilities to classify main types and types of them are shown in Table 5.2. The attributes shown here are the basis for our current investigations of the modelling process discussed below. Relief forms are relief units of homogeneous shape. The attributes are derived from plan and profile shape which can be described by different relief form indices.

Form elements	
Primary attributes	Plan and profile curvature
	Size order (scale of investigation)
Secondary attributes	Position in relation to the hierarchically higher- level unit
	Type of toposequence
	Neighbourhood relationship
	Height
	Distance to drainage divide
	Distance to drainage channel
	Height difference to drainage divide
	Height difference to drainage channel
	Shape
	Type and association of superimposed relief forms
	Subsurface material
	Geomorphodynamic processes
	Geomorphogenetic processes
	Geomorphochronology
Form facets	
Primary attributes	Gradient
	Aspect
	Plan and profile curvature
	Size order (scale of investigation)
Secondary attributes	See form elements

Table 5.2 Primary and secondary attributes for classifying form elements and form facets

An individual relief form can be synthesized by combining form elements. A mesoform may aggregate at a higher level in the model to mesoform associations or patterns which require a specific catalogue of spatially defined descriptors and a taxonomy of object relationships.

Typologically, relief forms are more flexible and more subjective. Their delimitation in the field survey often is pragmatical and can be influenced by the theoretical background of the scientist. For these reasons our systematic approach at present does not concentrate on the level of relief form definitions. Rather we are offering a catalogue of form facets and form

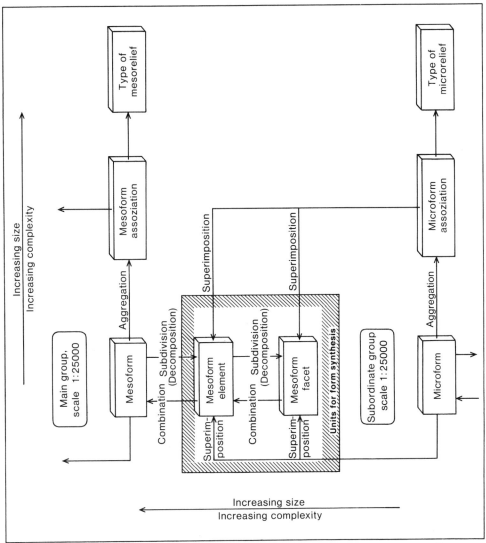

Figure 5.2 Systematization of complex geomorphological objects by analytical subdivision based on normalized components (design: D. Barsch and R. Dikau).

elements, which should allow a user-defined relief form synthesis (Dikau 1988b). In spite of these typological problems the automated modelling of highly complex relief forms and relief form patterns is an important aim of our future research. Descriptors for classifying these objects should relate to potential dynamics on hillslopes and their position within toposequences, as suggested, e.g., Speight (1974), and to neighbourhood relationships.

It is premature to predict to what extent a completely automated relief form analysis can be achieved applying the methods discussed. It can be said, however, that the methods provided by GIS are well suited for processing these problems.

Structure of the digital geomorphographical relief model (DGRM)

The primary aim of our current efforts is the transformation of the systematic approach described above into a computer-aided model. The following aspects formed the main objectives:

(1) The modelling has to generate form facets and form elements as basic relief units;
(2) The results have to serve as the basis for geomorphological and pedological mapping and for the simulation of geomorpho- dynamical processes;
(3) The modelling has to provide a systematic and methodological basis for the derivation of more complex relief units.

The relationship between model development and the requirements of geomorphological mapping as well as first model results have been discussed by Dikau (1988a) in more detail.

Source and quality of data

Our main data source for relief modelling are digital elevation models (DEM) available from the Ordnance Survey offices of the different federal states of the FRG. The grids were produced by orthophoto mapping and/or by the interpolation of digitized contours of the Deutsche Grundkarte 1:5000 (German Basic Map 1:5000) or analogous profiles. Unfortunately no uniform grid mesh size exists for the FRG. At the moment grids of 12.5, 20, 40 and 50 m are available. The DEMs used at present are compiled in Table 5.3. The grids are related to the Gauss-Krüger coordinate system.

Office of Ordnance Survey	Data source	Grid mesh size(m)	Mean height errors (m)	Interpolation program	Reference
(1)Baden-Württemberg	Analogue profiles and orthophoto mapping	50	±1.5 to ±5.0	SCOP	Sigle (1985)
(2)Rheinland-Pfalz	Contours (DGK5) and orthophoto mapping	40 20	±2.0 to ±9.0 ±0.5 to ±5.0	HIFI	Loskant (1985)
(3)Nieder-sachsen	Contours (DGK5) and orthophoto mapping	12.5	±0.1 to ±0.5	TOPSY	Staufenbiel (1980)

DGK 5 = Deutsche Grundkarte 1:5000 (German Basic Map 1:5000)
(1) = Use permitted under reference number: 4.49/28
(2) = Use permitted under reference number: 2.3468
(3) = Use permitted under reference number: B2a-0283/1.1

Table 5.3 Parameters of the digital elevation models currently used for relief modelling.

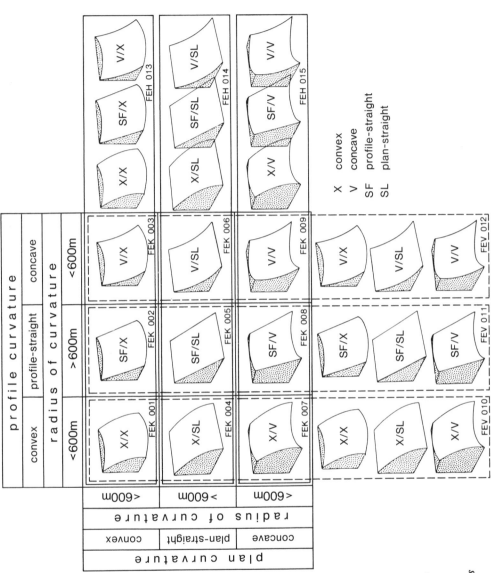

Figure 5.3 Classification of form elements by plan and profile curvature. Main type 1 is defined by a radius class of 600 m.

The accuracy and degree of generalization of the modelling is influenced significantly by height errors and spatial resolution of the DEMs, as has been discussed, e.g., by Mark (1984) and Douglas (1986) in relation to landforms. For the purpose of estimating the size of these influences a comparison of the modelling results with the corresponding geomorphological maps and their test in the field form an integral part of our researches. On the general assumption that the smallest object to be generated has at least twice the size of the grid, the mesh sizes listed in Table 3 correspond to the middle and upper range of the micro-relief, or, in relation to a scale of 1:25000, to map widths of 0.5 to 2 mm and in relation to a scale of 1:5000 to those of 2.5 to 10 mm.

The investigations have shown that 40 and 50 m grids produce no adequate results in the modelling of microforms. This means that the microform modelling cannot be reproduced with this degree of generalization of the DEM. In the scale used (1:25000) it is therefore necessary to perform a separate detailed geomorphological map of microforms such as gullies, erosion rills and strongly curved mesorelief elements (steps and breaks of slope).

Data structure

Following the terminology of Burrough (1987), the relief modelling is based on raster data structures and processing. Points, lines and areas found in maps are digitized as vectors and are converted into the compatible structure. Map overlays and object searches are raster-based.

The modelling process

Conceptually the modelling of the land surface consists of two parts together refered to as the digital geomorphographical relief model (DGRM) (see Figure 5.4). In the first part we derive the attributes relevant to relief modelling process as images and polygons. Furthermore this first section comprises routines for object generation and profile analysis. In the second part, which could be called the geomorphographical relief analysis the automatically extracted model results are subjected to extensive visual, empirical and statistical tests. The model is consists of several independent modules which will be discussed further. It is programmed in FORTRAN 77 and is currently running on a IBM 3090–180 under MVS/XA operating system at the University of Heidelberg.

Digital relief model

The generation of model attributes as raster images for 28 layers and as drainage channel and divide networks based on DEMs takes place in the model kernel digital relief model (DRM). Attributes 4-18 of Table 5.4 have been derived by modules developed at the Department of Physical Geography and Landscape Ecology, Technical University of Braunschweig. The program package is based on the strategy of analysing a 3 x 3 submatrix, described by, e.g., Evans (1980) and Zevenbergen & Thorne (1987). The modules furthermore can select drainage basins above variable and user-defined grid points. A detailed description of the algorithms is given by Bauer et al. (1985) and Bork & Rohdenburg (1986). For the derivation of attributes 19-30 the DRM was extended at the Department of Geography, University of Heidelberg (see Table 5.4).

The procedures are controlled by the two parameters ISEE and Emin:

ISEE: Optional regulation of the elimination of local minima below a variable size by a flooding simulation.

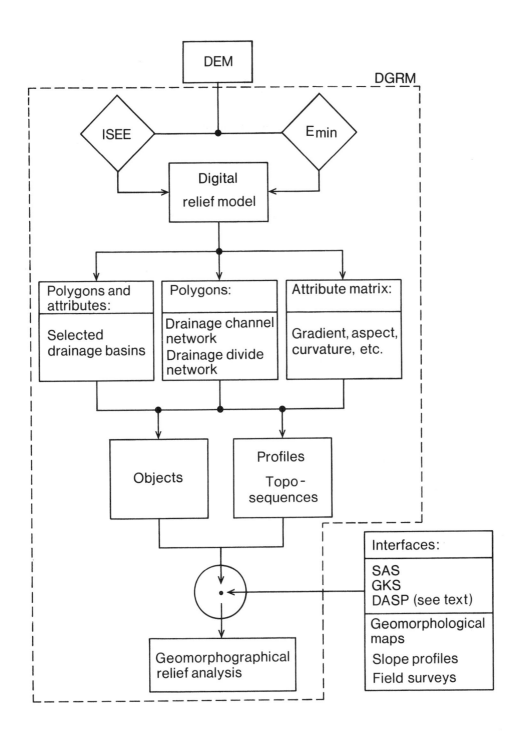

Figure 5.4 Structure of digital geomorphographical relief model (DGRM).

Emin: Optional regulation of the generation of drainage channel networks. A given point is incorporated into the network when its drainage basin exceeds this user-defined parameter.

Both these parameters permit to influence the simulation related to the geomorphological and hydrological characterization of the area investigated. Thus, closed depressions, as e.g. lakes or dolines, can be reproduced and the simulation results can also be adopted to the network density of real river systems. Furthermore the DEM can be smoothed in order to reduce inaccuracies of the data source or to apply the modelling to a given problem.

Key	Attributes (pixel level)	Attribute names	Unit of measure
A1	Gauss-Krüger x-coordinate	GKX	(m)
A2	Gauss-Krüger y-coordinate	GKY	(m)
A3	Altitude	ALT	(m)
A4	Aspect	ASP	(°)
A5	Slope gradient	GRA	(°)
A6	Radius of profile curvature	RCPR	(m)
A7	Radius of plan curvature	RCPL	(m)
A8	Size of drainage basin above surface point	SDBA	(m^2)
A9	Mean slope gradient of drainage basin above surface point	GDBA	(°)
A10	Distance to drainage channel	DDC	(m)
A11	X coordinate of related channel point	GKXDDC	(m)
A12	Y coordinate of related channel point	GKYDDC	(m)
A13	Shortest distance to drainage divide	SDD	(m)
A14	X coordinate of related drainage divide point	GKXSDD	(m)
A15	Y coordinate of related drainage divide point	GKYSDD	(m)
A16	Longest distance to drainage divide	LDD	(m)
A17	X coordinate of related drainage divide point	GKXLDD	(m)
A18	Y coordinate of related drainage divide point	GKYLDD	(m)
A19	Distance from channel to divide (short)	SDDC	(m)
A20	Distance from channel to divide (long)	LDDC	(m)
A21	Relative position to channel and divide (short)	PORS	(%)
A22	Relative position to channel and divide (long)	PORL	(%)
A23	Altitude of drainage channel	DADC	(m)
A24	Height difference to drainage channel	ALDC	(m)
A25	Altitude of drainage divide (short)	ALSDD	(m)
A26	Height difference to drainage divide (short)	DASDD	(m)
A27	Altitude of drainage divide (long)	ALLDD	(m)
A28	Height difference to drainage divide (long)	DALDD	(m)
A29	Real surface area of units	RAREA	(m^2)
A30	No. of drainage basin to which grid element belongs	NDBA	
	Attributes (object level)		
01	Length of boundary		(m)
02	Number of neighbours		
03	Length of common boundaries		(m)
04	Attributes of neighbours		
05	Statistical parameters of neighbour attributes		

Table 5.4 Attributes of the digital geomorphographical relief model (DGRM).

Classification and overlay

The classification of discrete attribute values on each data layer are processed by module REFAGN, the method used being variable. With module REKLA user-defined classes can be integrated into a catalogue. At the same time module REFAGN allows overlay procedures

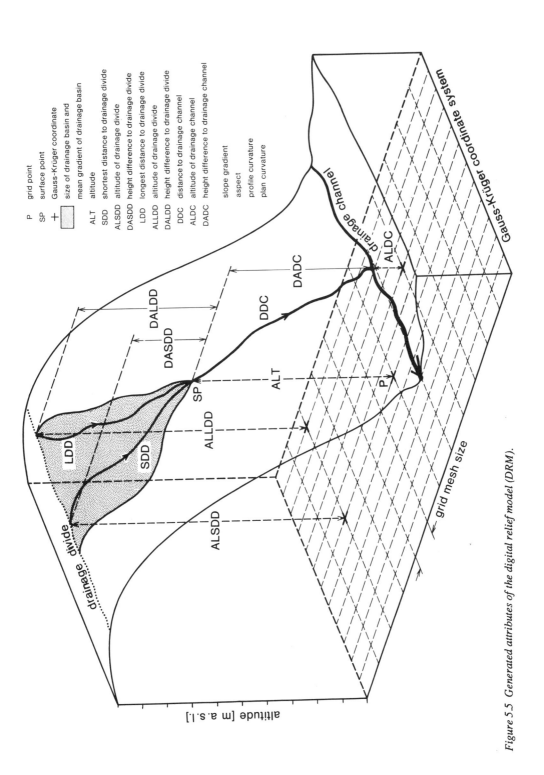

Figure 5.5 Generated attributes of the digital relief model (DRM).

of classified data layers. An interface to the central GKS plot program REPLOT permits the graphical representation of individual and combined layers.

Generation of individual objects and neighbourhood analysis

An individual object is defined as a region with uniform attribute values on a given data layer. The generation of objects takes place in module AGGREG following an algorithm of region growing by connected component labelling. Object attributes (see Table 5.2) are derived by different methods. The neighbourhood relationships, as the number of neighbours, attributes of neighbours, length of common boundary etc. is generated by AGGREG. An interface to the program package SAS (Statistical Analysis System) produces a second set of object attributes. This includes descriptive and spatial parameters derived from the unclassified attributes of each object. In this way a catalogue can be produced from which a relevant object type can be chosen and further processed. Their management, retrieval and selection is accomplished by the data management system DASP (Kühne 1983) of the Geological Survey of Lower Saxony (NLfB), and their graphical representation by REPLOT.

Geomorphographical relief analysis

The geomorphographical relief analysis (Figure 5.4) forms the second part of the modelling approach. Basically it consists of visual, empirical and statistical tests, including:
(1) statistical procedures supplied by the SAS package;
(2) overlay with further geoscientific maps (e.g. subsurface material);
(3) a parametrical relief classification by object selection; and
(4) comparison with results produced by other models (e.g. Jenson & Domingue 1988).

Parameterization of relief units

The application of objective, geometrical and reproducible attributes and rules to the definition and analysis of the land surface and its hierarchically organized components here is called formalization or parameterization. The process of parameterization not only requires the object analysis described above but also formalization of object affinities on different taxonomical layers. Finally, the procedures allow first attempts for aggregation or generalization of more complex objects from less complex ones. In the context of geometrical relief form description and geomorphometrical mapping projects a variety of attributes have been suggested and described by Speight (1968, 1974), Mark (1975), Evans (1980) and Pike (1988). These requirements make the geomorphographical formalization a task for techniques of geographical information systems. For the formal description of geomorphographical objects as relief forms, form facets or slopes, the basic principles of
(1) an adequate scale of investigation;
(2) a selection of geometrical and topological attributes adapted to this scale; and
(3) a DEM of suitable grid mesh size (if it can be chosen)
have to be defined. Their adaption is discussed in the following case studies.

Case studies

The taxonomy introduced above and its computer-aided transformation accomplished so far has been tested in different regions of the FRG. For two of them geomorphological maps at a scale of 1:25000 are available (Figure 5.6). The applications comprise:

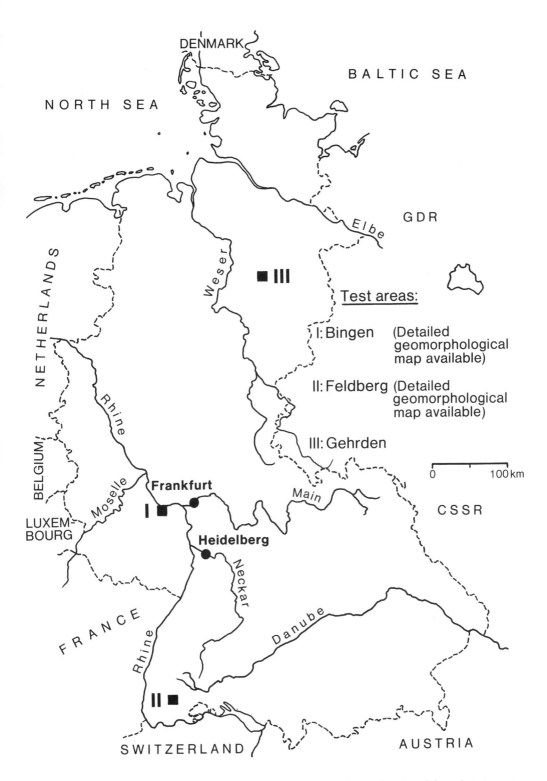

Figure 5.6 Schematic map of the Federal Republic of Germany, showing the location of the main test areas.

(1) statistical analysis of model attributes for problems of 'general geomorphometry', as frequency or hypsometrical distributions or correlations;

(2) extraction of relief units by overlay and object generation;

(3) analysis and statistics on the layer of 'specific geomorphometry' of objects;

(4) analysis of their neighbourhood relationships;

(5) profile analysis;

(6) production of computer maps (gradient, curvature, form element maps, etc.);

(7) empirical field test;

(8) feedback and, if necessary, modification of classification method and typology.

For field surveys a catalogue of terms and their corresponding symbols, a geomorphological classification key, must be provided. For this purpose we have used a system, which can be extended and which contains the form element catalogue of the GMK 25 (Dikau 1988b). The test areas of this project are shown in Figure 5.6:

(1) Test area I: GMK 25, sheet 21, Feldberg 1:25000, glaciation and periglacial conditions of the Central Upland of the Black Forest (Metz 1985);

(2) Test area II: GMK 25, sheet 11, Bingen 1:25000, southern part of the Rhenish Massif and its foreland, uplift situation and antecedent river (Andres et al. 1983);

(3) Test area III: topographical map 3623, sheet Gehrden 1:25000, loess-covered relief and hogbacks on the North German Plain, south west of Hannover.

The geomorphological situation here is dominated by the antecedent incision of the Rhine in its northern parts, a cuesta scarp region of the Rheinhessisches Plateau made up of tertiary clays and marls in its south eastern part, the Nahe river valley and the eastern section of the Saar-Nahe depression in the south west, and the inselberg region of the Rochusberg near the Nahe and the Rhine river junction. The following case studies are from 3 different areas of the DEM seen in Figure 5.7.

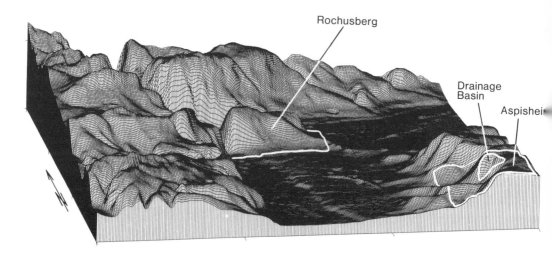

Figure 5.7 DEM for (grid: 40 m, 132 sqkm) of the GMK 25 sheet 11, Bingen, showing the three test areas discussed.

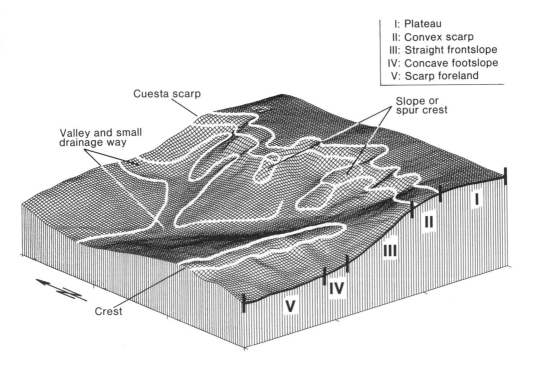

Figure 5.8 Digital elevation model (grid: 20 m, 4 km^2) of the Bingen/Aspisheim test area, showing schematically typical form elements of the cuesta scarp region.

Analytical subdivision

The principles of computer-aided analytical subdivision of a complex relief form are demonstrated for the example of the cuesta scarp in the Bingen/Aspisheim test area. The characteristic form elements of the scarp, which has been chosen from the geomorphological map and by field survey are shown schematically in Figure 5.8. It is now necessary to generate reproducable, geometrically unambiguous and geomorphologically meaningful relief units and to assess the methodological limits of the modelling procedure. Figures 5.9, 5.11 and 5.12 show the results based on a 20 m elevation grid. A frequency analysis of all form elements of main type 1 is contained in Table 5.5 and visualized schematically in Figure 5.3.

The convex cuesta scarp and the crest in front of it can be adequately modelled with the type FEV010 based on convex profile curvature, classified by a radius of < 600 m. As could be expected, this form element covers only 6.6% (8-connected) of the area. The small average size of objects is reflected in their large number. Their proportion amounts to 39% (8-connected). The crest and the cuesta scarp could be discriminated by a statistical analysis of secondary object attributes (Figure 5.10). In their object-related means of gradient and size of drainage basin both form element types occupy a characteristic range. This result is explicable in terms of their different location in space and their position in relation to the drainage divides. Of course, the range of values depends on the degree of scarp smoothness. This restricts the application of these ranges of values to other types of scarps. However, in summarizing the results it seems likely that the mean gradient and the mean of drainage basin size could be convenient general descriptors of cuesta scarps.

A second type of crest elements can be generated by convex plan curvature (Figure 5.11). They can be regarded as divergent slope crest or spur crest elements (FEH013) and cover almost half (41%) of the test area. As in areas of low slope gradients numerical uncertainties become evident the crest elements discussed above cannot be formalized with type FEH013. This type is used mainly in the subdivision of slopes.

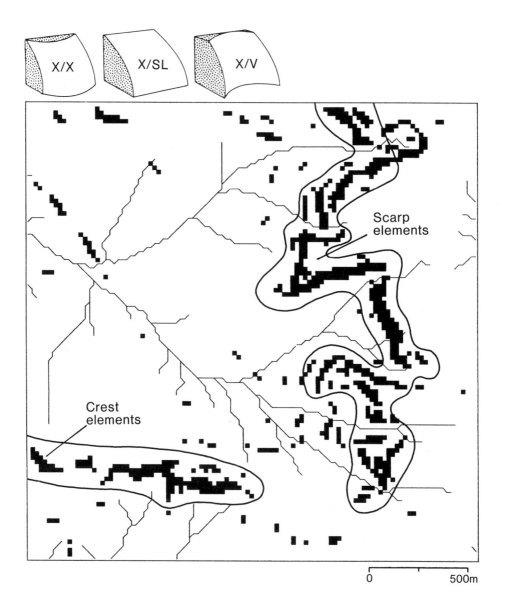

Figure 5.9 Form element map (crest, scarp), based on convex plan curvature, with an overlay of the generated drainage channels. Main type: FEH01, type: FEV010, grid: 20 m, radius of curvature: 600 m, ISEE: 0.2 km^2, Emin: 10000 m^2, smoothing level: 0.

Types of form elements		Test area Bingen/Aspisheim Area: 4 km² grid: 20 m Radius of curvature: 600 m			Test area Bingen/Rochusberg Area: 6.7 km² grid: 40 m Radius of curvature: 1200 m		
(Main type 1: FEH01)		Frequency distribution (%)			Frequency distribution (%)		
Key			Objects	Objects		Objects	Objects
Formal	Geometric	Area	8-connected	4-connected	Area	8-connected	4-connected
FEK001	X/X	4.8	7.6	7.6	3.2	3.7	3.7
FEK002	SF/X	35.7	12.0	16.2	37.1	18.1	18.0
FEK003	V/X	0.7	3.6	2.7	1.0	5.2	4.0
FEK004	X/SL	0.8	4.8	3.7	1.2	3.7	3.8
FEK005	SF/SL	11.4	41.5	38.3	14.3	29.8	32.4
FEK006	V/SL	0.8	5.1	3.8	2.3	3.5	4.0
FEK007	X/V	1.0	3.6	2.9	0.6	2.4	2.1
FEK008	SF/V	37.1	12.7	16.0	30.9	27.0	25.4
FEK009	V/V	4.1	9.2	8.7	3.1	6.3	6.3
FEV010	X/X X/SL X/V	6.6	39.7	36.5	5.0	20.8	21.9
FEV011	SF/X, SF/SL, SF/V.	84.0	4.5	15.4	82.3	25.0	31.3
FEV012	V/X, V/SL, V/V.	5.6	55.3	47.8	6.4	50.0	43.8
FEH013	X/X, SF/X, V/X.	41.2	11.2	15.2	41.3	20.0	19.4
FEH014	X/SL, SF/SL, V/SL.	13.0	75.0	67.1	17.8	40.2	48.5
FEH015	X/V, SF/V, V/V.	42.1	13.6	17.6	34.6	33.6	30.9

Table 5.5 Frequency distribution of form elements based on individual pixels (area) and number of objects. Test areas Bingen/Aspisheim and Bingen/Rochusberg. The types FEV010 to FEH015 are combined from types FEK001 to FEK009 (see Figure 5.3).

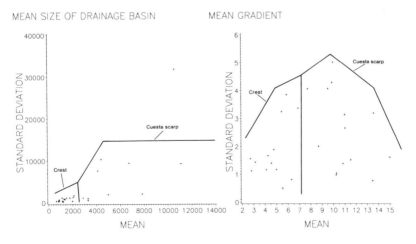

Figure 5.10 Discrimination of crest and scarp form elements by mean gradient (o) and mean size of the drainage basin (m²) of individual objects.

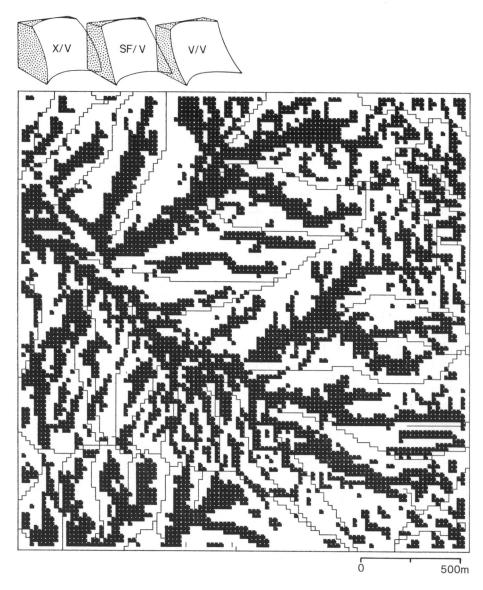

Figure 5.11 Form element map (slope crest, spur crest) based on convex plan curvature, with an overlay of the generated drainage divide. Main type: FEH01, type: FEH013 (model parameters– see Figure 5.9).

Valleys, small drainage ways and slope incisions can be modelled under similar limitations with the convergent form element type FEH015 (Figure 5.12). Today they represent the most active part of the cuesta scarp in which linear erosion processes dominate. The limitations imply that no satisfactory results can be expected from that area of the scarp plateau with low gradients (eastern part of the map in figure 12), which is characterized by wide and shallow depressions. The adequate results have been obtained by simulation of a valley floor inundation (case studies 3). The parameterization of the cuesta scarp form elements modelled for the Bingen/Aspisheim test area are summarized as follows (attribute names see Table 5.4).

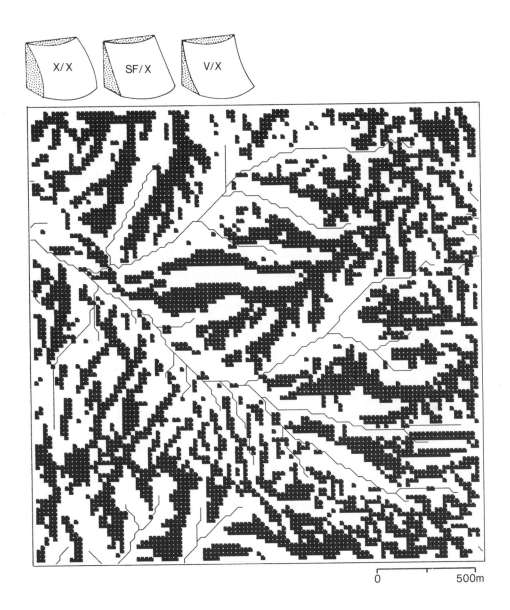

Figure 5.12 Form element map (valley, small drainage way) based on concave plan curvature, with an overlay of the generated drainage channels. Main type: FEH01, type: FEH015 (model parameter see Figure 5.9).

Model	Grid	= 20 m
	Smoothing Level	= 0
	ISEE	= 0.2 km^2
	Emin	= 10000 m^2
	Object generation	= 8-connected
Crest	RCPR/CONVEX (pixel level)	< 600
	GRA/MEAN (object level)	< 7
	SDBA/MEAN (object level)	< 2300
Scarp	RCPR/CONVEX (pixel level)	< 600
	GRA/MEAN (object level)	> 7
	SDBA/MEAN (object level)	> 2300
	and	< 14000
Valley and small drainage way	RCPL/CONCAVE (pixel level)	< 600
	GRA (pixel level)	> 2
Slope crest and spur crest	RCPL/CONVEX (pixel level)	< 600
	GRA (pixel level)	> 2)

Approaches to relief form synthesis

In a second case study, methods and relief attributes, needed in the synthesis of more highly complex objects, are introduced. Dependent on the model developed so far we have two strategies. On the one hand, a relief form can be analysed based on outline polygons which have been digitized from a geomorphological or topographical map. This can be, e.g., the boundary of a valley or a cuesta scarp. In this case the model will analyse the relief form basically on the form facet and element level. On the other hand, we try to generate more complex relief forms on the object level by applying the model attributes discussed and from neighbourhood relations. As part of this approach a catalogue of rules is compiled in order to make possible a taxonomy of relief forms based on object geometry and topology. This part of the project is in progress.

A form element map of the test area Bingen/Aspisheim (Figure 5.13) shows a comparatively generalized reproduction of the geomorphological mapping results at a scale of 1:25000. The 40 m elevation grid applied here does not produce the same model quality as a 20 m grid. In consequence only parts of the form element contained in the map could be modelled adequately. For this reason the grid was smoothed linearly (smoothing level= 1) and the form elements were generated by a higher class limit of curvature radius (<1200 m). The relief form 'Rochusberg' is composed of 4 types of form elements. These are the central crest, the hillslopes, the footslopes and the plains. Formal parameterization of form elements was accomplished as follows:

Model	Grid	= 40m
	Smoothing Level	= 1 (linear)
	ISEE	= 0.8 km^2
	Emin	= 10000 m^2
Crest	RCPR/ CONVEX (pixel level)	< 1200
Footslope	RCPR/ CONCAVE (pixel level)	< 1200
Hillslope	RCPR/ STRAIGHT (pixel level)	> 1200
	GRA (pixel level)	> 4
Plain	RCPR/ STRAIGHT (pixel level)	> 1200
	GRA (pixel level)	< 4

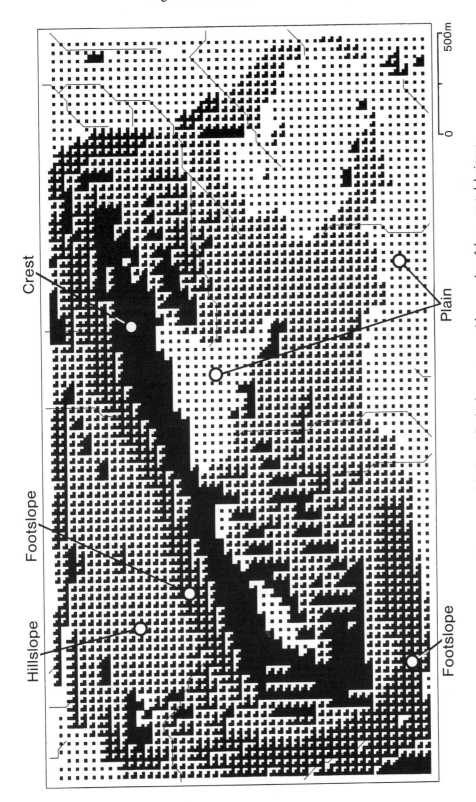

Figure 5.13 Form element map of the Bingen/Rochusberg test area with an overlay of the generated drainage channels. Grid: 40 m, radius of curvature: 1200 m, ISEE: 0.8 km^2, Emin: 10000 m^2, smoothing level: 1.

Model	Grid	= 40m
	Smoothing Level	= 1 (linear)
	ISEE	= 0.8 km^2
	Emin	= 10000 m^2
Crest	RCPR/ CONVEX (pixel level)	< 1200
Footslope	RCPR/ CONCAVE (pixel level)	< 1200
Hillslope	RCPR/ STRAIGHT (pixel level)	> 1200
	GRA (pixel level)	> 4
Plain	RCPR/ STRAIGHT (pixel level)	> 1200
	GRA (pixel level)	< 4

Dependent on the use of a larger radius a higher level of abstraction is obtained. In Table 5.5 this is recognizable in a more even distribution both at the level of areas and objects.

The synthesis of the specific relief form 'Rochusberg' is based on semi-automated processing of the neighbourhood relations. For this purpose an initial form element object is selected visually within the relief form under investigation. The generation of objects can be formalized as follows:

Model	Grid	= 40m
	Smoothing level	= 1 (linear)
	ISEE	= 0.8 km^2
	Emin	= 10000 m^2
	Object generation	= 8-connected
Relief form 'Rochusberg'	INITIAL FORMULATION	= object 27
	NEIGHBOURS (AREA)	> 0 <300,000
	COMMON BOUNDARY	> 80

The map in Figure 5.14 shows the result of the modelling as an association of form elements.

Classification based on drainage basin attributes

In a third case study the application of drainage basin attributes to the geomorphographical relief analysis is investigated. These are mainly the attributes which describe the local geometry as, e.g., the distances to the drainage channels and divides and the size of drainage basins above each surface point. Further types describe their allocation to selected drainage basin individuals (attribute A30, Table 5.4) or the length of the drainage channels into which a potential surface runoff of the form element would occur (Figure 5.15).

The height difference of surface points and the channel (attribute A24) permits the simulation of a valley floor inundation. Such a simulation separates the slopes from the neighbouring valley floors particularly in those cases where the radius of plan curvature obtains large values and in which the radius of the plan curvature shows inaccuracies and forbids a classification as discussed in case study 1. As the comparison of the DEM and the simulation of valley floors shows the shallow depressions of the cuesta scarp plateau in the eastern part of the plot can be better formalized by simulating an inundation (Figure 5.16). Based on height difference to drainage channel the valley floor is formalized as follows:

Model	Grid	= 20 m
	Smoothing Level	= 0
	ISEE	= 0.8 km^2
	Emin	= 10000 m^2
Valley floor	ALDC (pixel level)	< 1.0

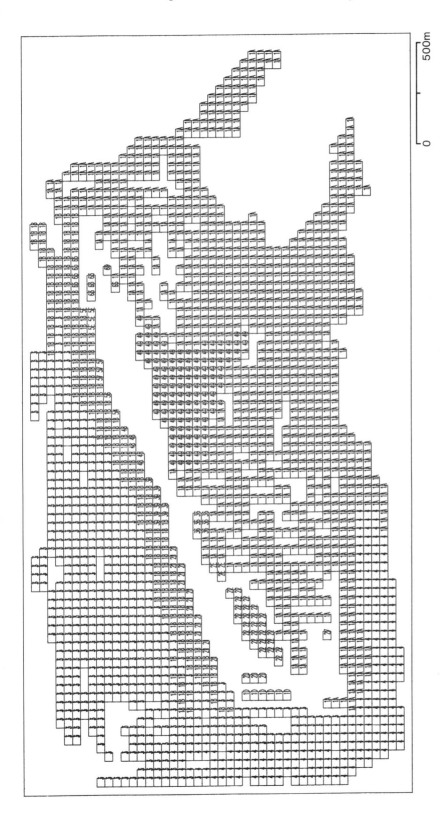

500m

0

Figure 5.14 Simple relief form synthesis (Rochusberg) by neighbourhood analysis of the central crest element (model parameters see Figure 5.13).

Although the examples discussed represent only a selection of the potential model application we hope to have demonstrated its suitability for research dependent on a geometrical and topological analysis of the land surface. In future we will concentrate on:

(1) the logical and spatial combination of the modelled objects with further geomorphological basic object types (e.g. subsurface material),
(2) the automated aggregation of single relief units to relief unit patterns, and
(3) the automated generalization of higher-level relief units from digital elevation models of high resolution.

Conclusions

From a Digital Elevation Model (DEM), we formally describe landforms through a hierarchical subdivision of the land surface into relief units. These units are defined by the logical combination of derivatives of the DEM and can then be combined semi-automatically to simulate the complex relief features seen in nature. Empirical testing of such simulations shows them to correspond well to natural landforms.

The quantitative analysis of landforms and their elements is an important basis for the investigation of the relationship between form and process in geomorphology. In addition, a quantitative geomorphological mapping system should provide objective classification of land surface features made up of definable components. Both aspects of relief analysis, the process-form interpretation and the classification of features, are important in the development of a computer-aided, raster- based model for relief analysis.

Our project is based on a relief-dividing model, digital geomorphographical relief model (DGRM) which requires:

(1) an objective classification and typology;
(2) defined components;
(3) a specific scale of investigation; and
(4) an adequate resolution of data sources.

From these, we derive a system for defining geomorphographical features and their spatial pattern. Starting from a DEM, we formalize the definition of geomorphographical objects by subdividing the land surface by using a taxonomical hierarchy of spatial relief units. In order of decreasing size and complexity, these units are (1) relief forms, (2) form elements, and (3) form facets. They are defined quantitatively as logical combinations of slope gradient, aspect, profile and plan curvature, distance to drainage divide, distance to drainage channel, elevation above the channel, gradient variability, etc. With this method, complex relief forms can be generated by combining form elements and form facets.

The procedure is currently being tested by case studies and empirical field surveys in a variety of terrains in West Germany: the Black Forest (Feldberg), a foreland of the Rhenish Massif (Bingen), and a loess-covered landscape in Lower Saxony (Gehrden). Automatic simulation of form elements and facets for these three areas gives spatial patterns that conform well to those defined by detailed geomorphological mapping (Geomorphologische Karte 1:25000 der Bundesrepublik Deutschland, GMK 25) but at considerably less expense. Relief form analysis in the Bingen test area suggests a classification of relief units at the level of elements according to plan and profile curvature. Depending on their vertical position, these units are systematically described by terms like "crest element", "slope element", or "valley element". In the same area, analysis of a 4 km^2 cuesta shows the situation of channels dissecting the scarp into a sequence of valleys and ridges with

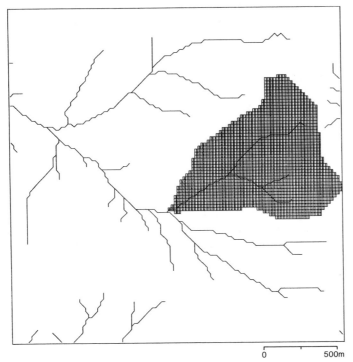

Figure 5.15 Selection of drainage basin objects for allocating form elements to drainage basin attributes; test area Bingen/Aspisheim (model parameters see Figure 5.9).

Figure 5.16 Parametrization of valley floors by simulating valley floor inundation (model parameters see Figure 5.9); inundation classes: 0 - 0.5 m, 0.5 - 1 m, 1 - 5 m, 5 - 10 m.

predominantly fluvial activity. Valley- side slopes and valley floors have also been investigated through their relationship between flood height and flooded area by simulating the condition of valley floor inundation.

Acknowledgements

The financial support for this project was provided by the Deutsche Forschungsgemeinschaft (German Research Foundation) under Ba 488/40-1-7 within the Priority Program 'Digitale Geowissenschaftliche Kartenwerke' (Digital Geoscientific Base Maps) (Vinken 1988). The author would like to thank the coordinator of the geomorphological sub-project Prof. Dr. Dietrich Barsch, and the members of our project for their help in preparing this paper. The component labeling routine was programmed by Dipl. Phys. Hanno Vieweger. Bernd Geiger developed the GKS program. Many thanks to Dr. Joachim Kadereit and to Prof. Nel Caine for help in translating.

References

Ahnert, F., 1988, Modelling landform change. In Anderson, M. G.(ed.) *Modelling Geomorphological Systems*, pp 375-400.

Andres, W., Kandler, O. and Preuss, J., 1983, Geomorphologische Karte der Bundesrepublik Deutschland 1:25000, GMK 25 Blatt **11**, 6013 Bingen.

Band, L. E., 1986, Topographic partition of watersheds with digital elevation models. *Water Resources Research*, **22**(1), 15-24.

Barsch, D., 1969, Studien zur Geomorphogenese des zentralen Berner Juras. *Basler Beitrage zur Geographie*, **9**, 221.

Barsch, D. and Liedke, H., 1980, Principles, scientific value and practical applicability of the geomorphological map of the Federal Republic of Germany at the scale of 1:25000, GMK 25, and 1:100000, GMK 100,. *Zeitschrift für Geomorphologie*, N.F., Suppl.-Bd.**36**, 296-313.

Barsch, D. and Liedke, H. (eds.), 1985, Geomorphological mapping in the Federal Republic of Germany. *Berliner Geographische Abhandlung* **39**, 1-94.

Bauer, J., Rohdenburg, H. and Bork,H -R., 1985, Ein Digitales Reliefmodell als Voraussetzung für ein deterministisches Modell der Wasser- und Stoff-Flüsse. *Landschaftsgenese und Landschaftsökologie*, **10**, 1-15.

Berry, J. K., 1987, Fundamental operations in computer-assisted map analysis. *International Journal of Geographical Information Systems*, **1**(2), 119-136.

Bork, H.-R. and Rohdenburg, H., 1986, Transferable parameterization methods for distributed hydrological and agroecological catchments models. *Catena*, **13**, 99-117.

Burrough, P.A., 1987, *Principles of Geographical Information Systems for Land Resources Assessment*. (Oxford: OUP).

Dikau, R., 1988a, Case studies in the development of derived geomorphic maps. In Vinken, R. Construction and display of geoscientific maps derived from databases, *Geologisches Jahrbuch*, **A104**, 329-338.

Dikau, R.,1988b, Entwurf einer geomorphographisch-analytischen Systematik von Reliefeinheiten. *Heidelberger Geographische Bausteine*, **5**, 1-46.

Douglas, D. H., 1986, Experiments to locate ridges and channels to create a new type of digital elevation model. *Cartographica*, **23**(4), 29-61.

Evans, I. S., 1980, An integrated system of terrain analysis and slope mapping. *Zeitschrift für Geomorphologie*, N.F., Suppl.-Bd. **36**, 274-295.

Jenson, S. K. and Domingue, J. O., 1988, Software tools to extract topographic structure from digital elevation data for geographic information system analysis. *Photogrammetric Engineering and Remote Sensing*, in press.

Kugler, H., 1974, *Das Georelief und seine kartographische Modellierung*. Diss. B, Martin-Luther-Universität Halle.

Kühne, K., 1983, DASP– Ein System zur Verwaltung und Auswertung geowissenschaftlicher Daten. *Geologische Jahrbuch*, **A 70**, 41-59.

Loskant, H.-J., 1985, Digitales Höhenmodell-Orthophoto-Luftbildkarte. Aufbau und Bearbeitungsstand. *Nachrichten- blatt Verw.- und Katasterverwaltung Rheinland-Pfalz*, Reihe 1, **28**(4), 266-282.

Mark, D. M., 1975, Geomorphometric parameters: a review and evaluation. *Geografisker Annaler*, **3** (4), 165-177.

Mark, D. M., 1984, Automated detection of drainage networks from digital elevation models. *Cartographica*, **21**, 168-178.

Metz, B.,1985, Geomorphologische Karte der Bundesrepublik Deutschland 1:25000, GMK 25 Blatt **21**, 8114 Feldberg.

Penck, A.,1894, *Morphologie der Erdoberfläche*. **1**.

Pike, R. J.,1988, The geometric signature: quantifying landslide-terrain types from digital elevation models. Mathematical Geology, **20** (5), 491-511.

Savigear, R. A.,1965, A technique of morphological mapping. Annals of the Association of American Geographers, **55**, 514-538.

Sigle, M.,1985, Das digitale Höhenmodell für das Land Baden-Württemberg. *Nachrichten der Karten- und Vermessungswesen*, **95**, Reihe 1, 143-154.

Speight, J. G., 1968, Parametric description of land form. In Stewart, G. A. (ed.) *Land evaluation*, pp. 239-250.

Speight, J. G., 1974, A parametric approach to landform regions. In Institute of British Geographers, Special Publication **7**, pp 213-230.

Speight, J. G., 1984, Landform. In McDonald et al. (eds.) *Australian soil and land survey field handbook*, pp 8-43.

Staufenbiel, W.,1980, Das topographische Datenbanksystem TOPSY. *Nachrichten der Karten- und Vermessungswesen*, Reihe 1, **81**, 101-110.

Vinken, R. (ed.), 1988, Construction and display of geoscientic maps derived from databases. *Geologisches Jahrbuch*, **A 104**.

Zevenbergen, L. W. and Thorne, C. R., 1987, Quantitative analysis of land surface topography. *Earth Surface Processes and landforms*, **12**, 47-56.

Chapter 6

Visualisation of digital terrain models: techniques and applications

Robin A. McLaren and Tom J. M. Kennie

Introduction

The representation of relief is a fundamental component of the cartographic process. A wide range of techniques of representing the topographic variations of the earth's surface on a two dimensional surface have been developed and these vary both in their symbolic content and in their degree of realism (Figure 6.1).

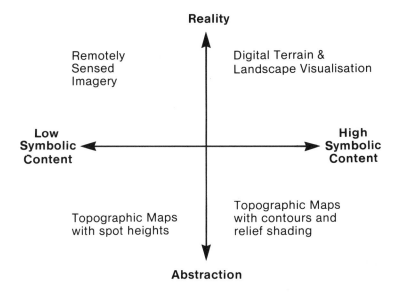

Figure 6.1 Classification of techniques for visualising terrain.

Recently, the use of digital terrain models (DTM) and computer graphics techniques have become a more common method not only for generating computer visualisations of the terrain but also for assessing the impact of manmade objects on the landscape. Thus in many cases the terrain model providing the geometric description of the earth's surface will be supplemented by descriptions of significant landscape features. Vegetation and cultural information as well as design objects will often be combined with the terrain model to create a scene with a higher degree of realism. Models of this form can be referred to as landscape models. In small scale applications, this landscape information will probably be polygonal land use data, while at larger scales it may include explicit three dimensional geometric descriptions of individual features or blocks of features. This paper therefore reviews the techniques used to create and enhance these digital terrain and landscape visualisations and considers some of their small and large scale applications.

There are many interrelated reasons which account for the growth of interest in this area of visualisation. Certainly one of the most significant has been the improvement in the price/performance ratio of modern computers. The display architectures and memory facilities available in the current generation of graphics workstations enable results to be created at a price which could not have been contemplated 10 years ago (Gelberg and Stephenson 1987).

The general improvements in the field of computer science have also had an immense impact on the design of the instrumentation used to acquire terrain data. Electronic tacheometers, analytical stereoplotters and automated cartographic digitisers are now widely used to produce grid or triangular terrain models. This ability to create DTMs at an economic cost has had an enormous effect on the use of DTMs generally for mapping and civil engineering design and has also created a demand for more sophisticated methods of displaying the data. Several organisations (mostly military) have also created small scale regional, national and international terrain databases. The provision of these terrain databases has also been a significant influence on the development of methods of visualising the terrain.

Concurrent with the improvements in computer hardware have been some extraordinary developments in the algorithms used for general scientific visualisation. The impetus for many of these software developments has come from the needs of the simulation, CAD/CAM, advertising and entertainment industries. The requirements of the film industry for computer generated animations have been particularly influential in this respect. The importance of this area of technology is indicated by the recent formation by the USA National Science Foundation of a group to examine user needs in the general field of Visualisation in Scientific Computing (McCormick et al. 1987).

While there is no shortage of techniques for general visualisation purposes, the characteristics of the earth's surface can limit the applicability of many of these techniques. Among the more important differences between terrain and landscape visualisation and other forms of visualisation are the following considerations.

(a) Natural phenomena are inherently more complex than man made objects and thus more difficult to model;

(b) The earth's surface is not geometric in character and cannot be modelled effectively by using higher order primitives such as those used in CAD/CAM applications;

(c) The terrain and landscape model dataset sizes are considerably larger than the datasets in many other forms of visualisation;

(d) There are generally higher constraints on geometric accuracy than in many other applications;

(e) The scenes are not spatially compact. Therefore, the modelling may involve multiple levels of detail based on the object/viewpoint relationship;

(f) The optical model is more complex due to the effects of atmospheric refraction and earth curvature which are encountered in extensive datasets.

Finally, user requirements have also influenced developments in this field. For example, as a consequence of recent changes in EEC contract regulations (European Economic Community, 1985), visual impact analysis is now a legal requirement of certain major construction projects. The needs of users in the field of geographic information systems (GIS) for display technologies to illustrate the results of sophisticated spatial analyses have also been a significant factor. Military users have also been a major driving force and a selected number of applications of terrain and landscape visualisation are presented in the latter half of the paper.

Creation of digital terrain models for visualisation

The acquisition of terrain data is a critical stage in the process of visualisation since it will have a considerable influence on the fidelity and hence the degree of realism which can subsequently be achieved. Four distinct approaches to the creation of DTMs for terrain and landscape visualisation can be identified:

(a) Ground surveying techniques using electronic tacheometers and data collectors for the production of high accuracy models of limited areas;
(b) Photogrammetric techniques using either analytical stereoplotters, possibly with correlators attached, or instruments for orthophotograph production for medium to high accuracy models of large areas;
(c) Graphics digitising techniques where contour lines on existing maps are transformed into digital coordinate data using either manual or automatic, line following digitisers or automatic, raster scan digitisers (in both cases the accuracy is significantly lower than the previous two techniques although the coverage can be very large); and
(d) Generation of simulated terrain using either computer based geomorphological models of the earth's surface or fractal geometry to generate more abstract surfaces.

Approaches to sampling real terrain

Regardless of the technique which is used to measure the terrain, the matter of determining the optimum number of points to acquire is a central issue, since the density of sampling has a direct influence on the accuracy of the terrain representation. Several theoretical models have been developed in recent years in an attempt to resolve this problem. The most commonly reported methods are those based on the analysis of either the power spectrum of the terrain using Fourier analysis (Tempfli 1980), variograms of the terrain (Frederiksen et al. 1985) or self similarity using fractal techniques (Muller and Saksono 1986). These theoretical techniques are currently limited in their applicability although they show some potential in the determination of sampling strategies for photogrammetric mapping over large areas (Balce 1987).

In practice it is possible to distinguish three commonly occurring approaches to data sampling: selective (irregularly located) sampling using both ground and photogrammetric survey techniques, systematic (regularly located) sampling using photogrammetric techniques and hybrid sampling using cartographic digitising.

Selective (irregular) sampling using ground surveying techniques

For high accuracy terrain representation over relatively small areas selective ground sampling methods are used to acquire data. In this case the surveyor chooses points at significant changes in terrain form for example, at the tops of hills, breaks of slope, ridges and so on. The outcome is eventually a series of irregularly located points representing the terrain. Surface representation is subsequently derived by using either planar or curvilinear triangular elements or by grid interpolation.

Systematic (regular) sampling using photogrammetric techniques

An alternative strategy is normally adopted when sampling is performed by close range or aerial photogrammetric techniques. In this case a systematic pattern of heights is measured, normally in a grid fashion. The primary advantage of this technique is that all, or part of the process can be automated, for example by using analytical stereoplotters to drive to pre-set positions and/or by using correlators to automate the height measurement. The well known disadvantage of this approach, that the distribution of the datapoints may not be related to the nature of the terrain, can be overcome by using the progressive and composite sampling techniques derived by Makarovic and reported more recently by Tempfli (1986).

Hybrid sampling using cartographic techniques

Digitising of existing maps leads to a different approach to sampling from those discussed in the previous two examples. For example, in the case of manual line following digitising, the sampling of the terrain often combines features of both the selective and systematic procedures outlined previously. The digitising of maps as an input to the Panacea software suite (McCullagh, 1988) involves initially forming an irregular triangulated data structure followed by the generation of a grid DTM from the triangulated data.

Accuracy of DTM's for visualisation

The relative accuracies of DTMs derived by ground survey, photogrammetry and digitising existing map sources are illustrated by Table 6.1. It is clear from this table that the fidelity of the elevations derived by different techniques can vary substantially. For small scale visualisation applications (such as flight simulation), these inaccuracies in the absolute values in the database may be relatively unimportant provided that the dataset is free from gross errors and the relative heights of points are within an acceptable tolerance. As long as these

Source	Data measured	Type	Accuracy factor
Ground survey Large scale model (1:500)	Directly measured spot heights	1	1
Photogrammetric measurement Aerial photographs for topographic mapping (e.g. 1:10,000)	Photogrammetric spot heights (measured in a stationary mode)	2	5
	Photogrammetric contours (measured in a dynamic mode)	3	15
Cartographic digitising Medium scale topographic maps (e.g. 1:50,000)	Contours generalised from medium scale maps	4	25
	Contours measured by field survey in the 19th century	5	50
Small scale topographic maps (eg.1:250,000)	Spot heights at grid nodes derived by interpolation from digitised contours	6	500

Table 6.1 Relative accuracies of DTM elevation values as derived by various surveying techniques (adapted from Petrie and Kennie 1988). (Accuracy factor 1 = high accuracy; 500 = low accuracy)

limitations in accuracy are recognised by and are acceptable to users lower quality data are tolerable.

For large scale applications in landscape architecture, civil engineering and opencast mining, however, the inaccuracies in elevation resulting from the data derived from small scale map sources described in Table 6.1 would generally be unacceptable. This would be particularly important if the terrain model was being used for visual impact analyses.

Approaches to creating simulated terrain

An alternative approach to the creation of terrain models for visualisation purposes is to produce a simulated surface rather than measure a real surface. Two distinct techniques, which have been used for such exercises, are those which simulate geomorphological processes and those which use fractal geometry.

Geomorphological modelling

Computer simulations of terrain attributes such as slope development (Armstrong, 1976) or of large scale erosional landforms (Craig, 1980) have attracted considerable interest in the geomorphological community in recent years. A novel application of several of these concepts to the creation of a synthesised terrain surface for a vehicle simulation exercise can be found in Griffin (1987). The aim of the project was to simulate a fluvial landscape by the establishment of a network of streams over an initial surface. The simulation was based on a primary surface (a horizontal/tilted plane consisting of a grid of spot heights) with different surface lithologies (to simulate resistance to erosion) and the movement of streams based on a random walk algorithm.

Durrant (1987) also discusses the use of simulation, in this case to assess the dynamic nature of the discharge from tailings dams. These dams are structures built from mine waste which are continuously constructed during their service lives. The paper considers methods of modelling the nature of the depositional cycle from the discharge points by simulating the flow pattern. The CIS Medusa system was used for these operations.

Most of the geomorphological models include both deterministic and stochastic (for example random walk) elements. Realistic simulated terrain has been also been produced solely by stochastic processes of which the most common involves the use of fractals.

Stochastic modelling using fractal geometry

The terms fractal and fractional Brownian motion were originally devised by Mandelbrot and Van Ness (1968) to denote a family of stochastic models which could be used to represent many spatial or temporal phenomena. Since then the concept has been applied by Mandelbrot and many other researchers to the process of terrain and landscape simulation (Mandelbrot, 1975; Fournier et al., 1982; Voss, 1985; Clarke, 1987).

One of the central themes of fractal geometry is the concept of self similarity. This property of an object describes the invariance of the object under changes of scale, or alternatively the degree to which objects appear similar, but nevertheless different, at varying levels of magnification. This idealised form of fractal geometry rarely, if ever, occurs in nature although statistical or stochastic self similarity is considered by many to be evident in real landscapes. Mandelbrot and others have shown that landscape form can be described by the fractal dimension of the surface, a non-integer value between 2 (a perfectly smooth surface) and 3 (an infinitely variable surface). Real landscapes have been found to have fractal dimensions between 2.1 and 2.3, although values as high as 2.75 have been reported.

The consequence of accepting self similarity is to suggest that the spatial structures inherent within the landscape repeat themselves at all scales. The evidence from several sources seems, however, to indicate that self similarity does not occur in nature. One explanation suggested by Roy et al. (1987) is that the earth forming processes have operated on the landscape for varying amounts of time and consequently the earth's surface manifests the scale at which these different processes operated. Burrough (1985) also confirms this opinion and suggests that "most landscapes are not the result of a single dominant process but are the result of the complex interaction and superimposition of many processes". In spite of these limitations, simulated landscapes generated with fractal dimensions between 2.2 to 2.3 have been found to produce realistic looking images and have been used in several animated sequences for science fiction films and advertising (Plate I).

Computer graphics techniques for displaying digital terrain and landscape visualisation

This section of the paper examines the various techniques which can be used to create computer visualisations of the terrain and associated landscape features. The degree of realism which can be achieved in such a process is dependent upon several factors including the nature of the application, the objective of the visualisation, the capabilities of the available software and hardware and the amount of detail recorded in the model of the scene. At one end of the realism spectrum are highly abstract, simple, monochrome wireframe models while at the opposite end are highly realistic, fully textured and coloured images. While the quest for realism ultimately attempts to produce images which are indistinguishable from reality, it is generally impractical at present to use such images in dynamic environments such as landscape design or avionics. For the future, however, it seems likely that there will be a trend towards the creation of more sophisticated dynamic imagery.

Assumed display and output devices

Raster scan displays have become the de facto technology for visualisation applications. Such devices are typically a 19 inch diagonal CRT screen with a typical resolution of 1024 x 1280 pixels and a capability of displaying up to 16.8 million colours. The scene generation approaches described in the paper will concentrate on techniques which rely on raster technology. These techniques will also continue to be applicable as the current CRT technology is superseded by new devices such as flat panel displays.

Despite this predominance of raster display technology in forming images for visualisation, the hard copy output is often produced by a vector plotter. For many applications the level of abstraction produced by such devices is sufficient for the scene analysis, especially at the planning and early design phases. Therefore, vector techniques will also be explored in describing scene generation approaches.

Methods of displaying DTM data

Some relatively simple techniques have been developed for displaying DTMs on a computer graphics display. More sophisticated approaches which involve the use of perspective transformations and rendering techniques are discussed subsequently.

Probably the most common method which is used to indicate height variations on a two dimensional surface is the contour. Traditional graphical maps normally portray contours in a single colour and use a variable line width to assist in the identification of major height variations. Colour graphics displays may also be used to portray contour lines, although colour variations are more commonly used to give an appreciation of the height variations rather than variable line widths. An extension of the colour coding of contour lines is to colour code the areas between contour lines as an aid to improving the appreciation of the topographic variations (Plate II). In both cases the choice of colours which are used to represent the lines or areas is of crucial importance. Often a single colour of varying hue and intensity can be particularly effective.

In addition to the display of contours it is possible to derive additional information from a DTM which can improve the visualisation of a particular attribute of the earth's surface. Thus it is quite common to derive the slope or gradient of the terrain and display this either in the form of slope vectors or as a colour coded image. Aspect or direction of maximum slope is a further attribute which can be derived from the terrain data and displayed as a colour coded image. In both cases the choice of colours can also have a major impact on the interpretability of the resulting images. By combining both slope and aspect information with the position of the sun as the illumination source, it is also possible to generate, in monochrome, the effect of hill shading or shaded relief (Plate III).

These techniques are capable of meeting the display needs of many users. They do, however, exhibit a number of deficiencies including the production of a highly abstract view of the terrain from a fixed viewing position with few facilities for including significant landscape features in the displayed image. The following sections describe methods of addressing these limitations.

Components of an information model for visualisation

The majority of computer graphics techniques have been developed to address the visualisation of objects that have their geometry defined using a mesh of planar surfaces such as triangles. Some algorithms are available to process higher order surfaces such as bicubics. However, these still tend to be the exception, forcing most approaches to decompose higher order surfaces into their component planar surfaces prior to graphic processing (Snyder & Barr, 1987).

These geometric descriptions of a surface can also be supplemented by parameters describing the surfaces' visual characteristics such as colour, reflectivity and texture that in effect clothe the terrain. The allocation of the correct colours to objects is crucial in achieving realism and the process is not as simple as it may initially appear. Difficulties can arise when attempting to select a computer graphic generated colour, on a screen or hard copy device, to match the colour of a natural object. In part, this is due to the "artificial" colour models used to present colour to the user of graphic systems. Schwartz et al. (1987) conducted a revealing experimental comparison of a number of different colour models encountered in the computer graphics world.

Ideally, a scene generation system should be an integral facility of a GIS where the objects composing a scene could be extracted from the database, together with their associated three dimensional graphical representations, to create a required scene.

Methods of rendering terrain and landscape models

Having assembled the geometric model of the terrain and landscape objects to be visualised, the next stage of visualisation is to render this three dimensional description of a scene onto a two dimensional display device.

As well as the model of the terrain and associated landscape features contained within the required scene, the rendering process requires a number of other parameters to be defined including:

(a) The viewing position and direction of view of the observer;

(b) A lighting model to describe the illumination conditions;

(c) A series of "conditional modifiers" (McAulay, 1988) parameters which describe the viewing condition of the landscape objects (under wet conditions for example, the surface characteristics of objects are quite different from dry conditions);

(d) A set of "environmental modifiers" (McAulay, 1988), parameters which describe atmospheric conditions and may model effects such as haze; and

(e) A sky and cloud model representing the prevailing conditions.

All, or part of the information above may then be used by the scene rendering process to generate a two dimensional array of intensities or pixel values that will be displayed on the raster display device. The complexity of the rendering process is directly dependent upon the degree of image realism required by the user. There is always a trade off between quality of image and performance/cost. For example, in the development of the Reyes rendering

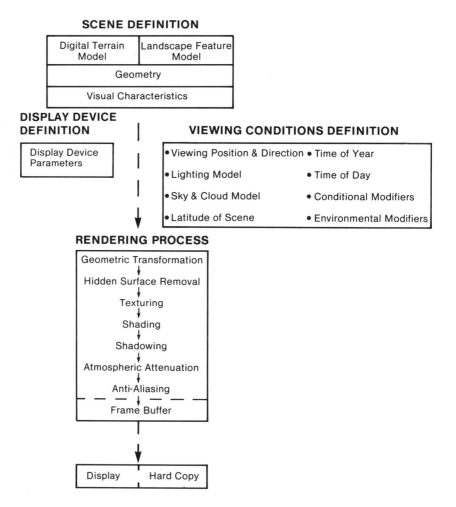

Figure 6.2. Overview of the rendering process.

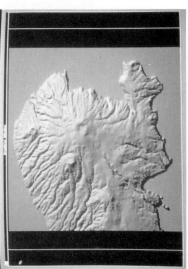

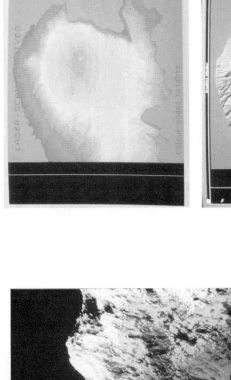

Plate I (above) Simulated fractal landscape with dimension of 2.2, R. Voss, IBM Ltd.

Plate II (right, upper) Area fill colour coded heights. Laser-Scan Laboratories Ltd.

Plate III (right, lower) Shaded relief for area in Plate II. Laser-Scan Laboratories Ltd.

Plates

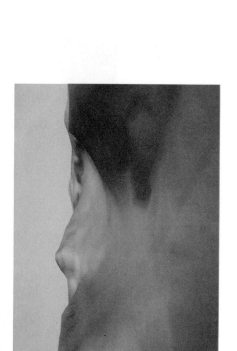

Plate IV (left, upper) Facetted shading of a landscape. Cambridge Interactive Systems Ltd.

Plate V (left, lower) Gourand shading of the same landscape as in Plate IV. Cambridge Interactive Systems Ltd.

Plate VI (above) Aeronautical chart of part of Alaska showing hill shading with colour coded height variations. USGS EROS Data Center.

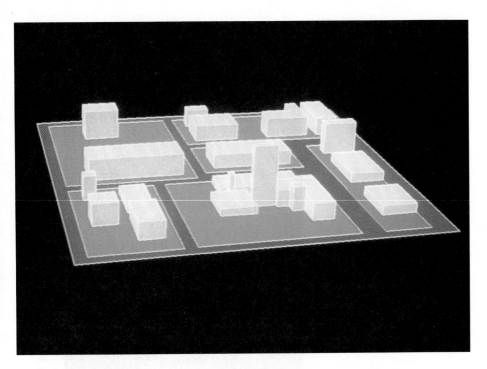

Plate VII A 3-dimensional urban map: representation of an industrial estate in a Dutch new town.

Plate VIII A digital elevation model: SO_2 pollution in the Athens area. Above the DTM, which is a combination of the height and land use information, a 2-dimensional isoline map of the SO_2 concentrations in the area is displayed.

Plates

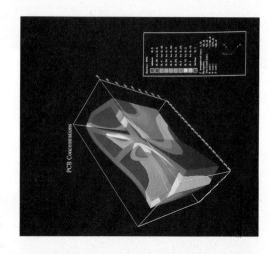

Plate XI Iso-surfaces cut by polygons.

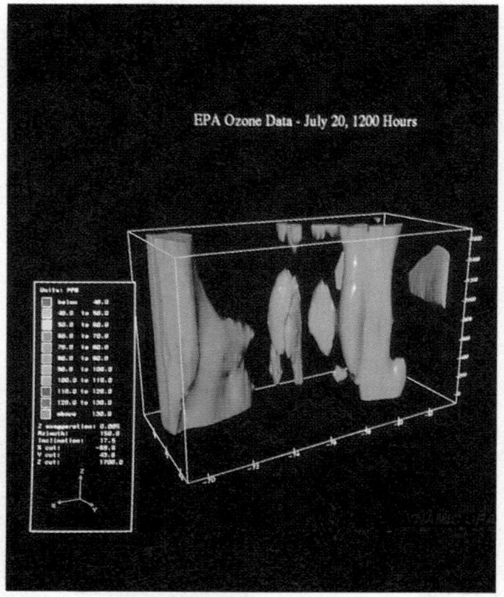

Plate X Example of an iso-surface.

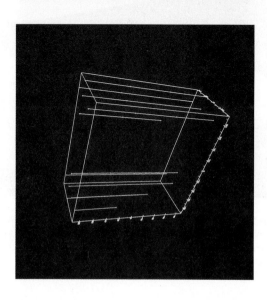

Plate IX 3-dimensional source data for PCB pollution.

Plates IX-XVII, Dynamic Graphics Inc.

Plates

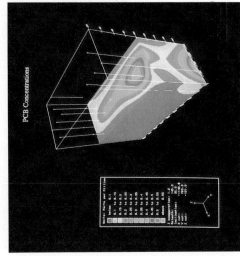

Plate XIV Sliced cube display (subsetting the block).

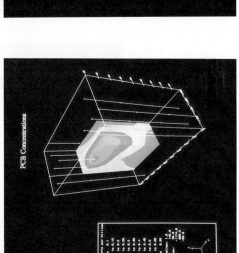

Plate XIII Sliced cube display (across full block).

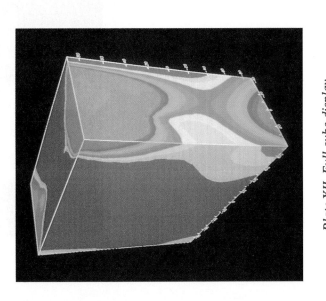

Plate XII Full cube display.

Plates

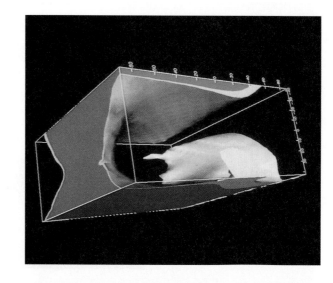

Plate XVII Concentrations below a value in the block.

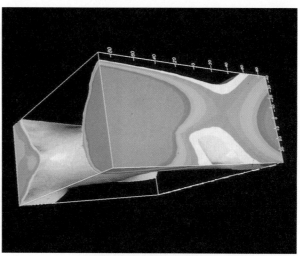

Plate XVI Concentrations above a value in the block.

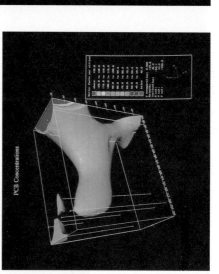

Plate XV Iso-surface display after slicing.

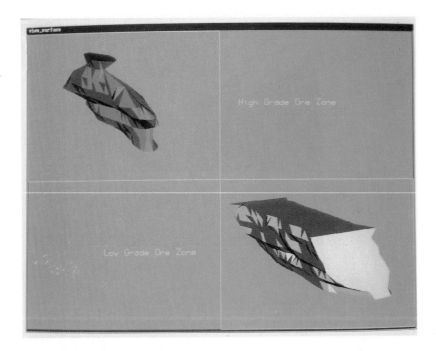

Plate XVIII Shaded display of boundary representation - model of an ore deposit. The high grade ore zone (green) is a subset of the low grade ore zone (yellow).

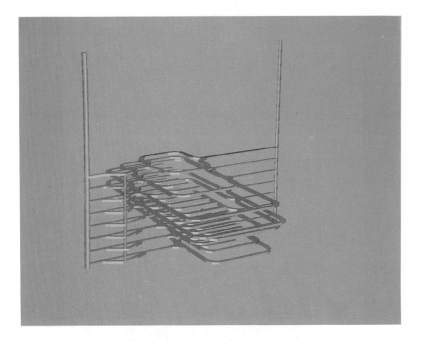

Plate XIX Shaded display of boundary representation - model of an underground mine layout.

Plate XXII Cut away view of a linear octree representation - model of a complete mining project. The model comprises a high grade ore zone (blue), low grade ore zone (yellow), mine layout (grey) and a mined out open pit (red).

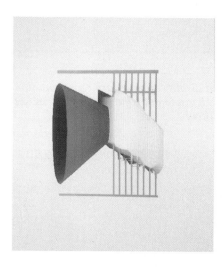

Plate XXI Shaded display of a linear octree representation - model of a complete mining project. The model comprises a high grade ore zone (which is not visible), a low grade ore zone (yellow), a mine layout (grey) and a mined out pit (red).

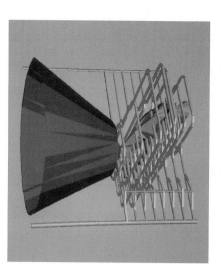

Plate XX Shaded display of boundary representation - model of a complete mining project. A graphical merge of a mine layout (grey), a high grade ore zone (green) and a mined out pit (red).

system (Cook et al., 1987) the performance objective was to create a two hour film in one year. At 24 frames per second, this is equivalent to rendering a frame in three minutes.

An overview of the rendering process is shown in Figure 6.2 and the remainder of this section briefly describes the steps involved in the scene rendering procedure. The sequence of steps may vary depending on the rendering approach chosen.

Geometric transformations

The three dimensional terrain and landscape information is normally mapped into two dimensional space by a perspective projection, where the size of an object in the image is scaled inversely as its distance from the viewer. An example of this form of projection is illustrated by Figure 6.3. In this class of projection, the interpretation of depth is usually based on the assumption that the largest of a class of objects is closest to the viewer and, in effect, it models a "pin hole" camera.

Occasionally, the isometric projection (Butland, 1979; Dubayah and Dozier 1986) is used to visualise terrain and the modelling of sub-surfaces. This is an axonometric projection that has the useful property that all three principal axes are equally foreshortened, providing equal scale in all three directions. In addition, the principal axes are projected so that they make equal angles with one another. This is useful for some forms of analyses.

For applications where the geometric fidelity of the rendered scene is of vital importance, for example the creation of a photomontage product in visual impact assessment, it is necessary to incorporate both earth curvature and atmospheric refraction corrections into the viewing model.

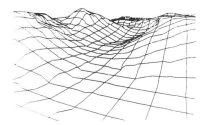 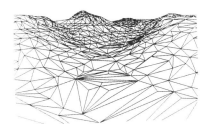

Figure 6.3. Depth cueing using (a) gridded structure and (b) triangular structure. Cambridge Interactive Systems Ltd.

Depth cueing

When a three dimensional scene is rendered into two dimensional space with any level of abstraction, an ambiguous image will probably be portrayed. This is compounded by the fact that our eyes are not a window into the world, but instead the world is created in our mind based on preconceived models that vary from person to person (Gregory, 1977). Therefore, if new computer graphic presentation concepts do not match these preconceived models, then they are open to mis-interpretation. A wire frame model presents the viewer with the maximum degree of ambiguity. To compensate for this loss of inherent three dimensional information, techniques have been developed to increase the three dimensional interpretability of the scene using depth cueing techniques that attempt to match the perceived computer generated image to our "natural" visual cue models.

Hidden edge removal

At the highest level of abstraction, objects can be modelled by their edge components, creating a three dimensional skeleton form called a wire-frame model. When this is subsequently viewed as a perspective projection, the result is confusing and ambiguous due to edges being visible that would normally be hidden if a solid model had been used. The interpretability of depth relationships of the image can be improved by removing the hidden edges from the image. The approach to hidden edge removal is equivalent to hidden surface removal which is described in the following section.

Although images formed by wire-frame models with hidden edge removal are primitive images with no pretence at being realistic, they still portray form and geometric fidelity and provide an inexpensive technique for visualisation. Major advantages are the low overheads in their production and their ability to be output on standard vector plotting devices. The visualisation of terrain models in wire-frame form is not limited to the direct output of the grid or triangular network; ridge lines and other visual forces, derived from the DTM, may be used discreetly or in combination with the DTM structure lines.

Hidden surface removal

An extension of the wireframe display is to 'fill' the component surfaces defining the objects, giving the impression of solid objects. Just as edges were removed in the context of wireframe models to eliminate ambiguities, hidden surface removal techniques are employed to remove those edges and surfaces that are obscured by other visible surfaces. The hidden surface capability provides a further step forward in the quest for realism.

The implementation of the technique of hidden surface removal is computationally expensive, especially for complex landscape scenes, where the rendering process can involve hundreds of thousands of surfaces. Therefore, the challenge has encouraged a wide variation of algorithms. These can be categorised into two fundamental approaches: image space and object space algorithms. One of the simpler and more popular object space algorithms is the depth sort algorithm (Newman and Sproull, 1979; Sechrest and Greenberg, 1981), in which the polygons to be displayed are sorted by their distance from the viewpoint and scan converted into the refresh buffer of the display device in order of decreasing distance. Since the nearest polygons are placed into the refresh buffer last, they obscure the polygons that are further away by overwriting them in the refresh buffer. The priority for displaying the polygons cannot be simply based on a single maximum Z value since depth ambiguities may occur, which have to be resolved. Although this algorithm uses the structure of the display device to aid in the surface removal and is conceptually simple, it does have the limitation of requiring the polygons to be sorted prior to display.

One of the most popular algorithms used for the elimination of surfaces in a landscape scene is the Z-buffer or refresh buffer image space algorithm. This approach assumes that the display device has a Z buffer, in which Z values for each of the pixels can be stored, as well as the frame buffer in which the intensity levels of the pixels are stored. With the reduced cost of memory, this is becoming a standard facility of display devices. Each polygon is scan converted into the frame buffer where its depth is compared with the Z value of the current pixel in the Z buffer. If the polygon's depth is less than the Z buffer value then the pixel's frame buffer and Z buffer values are updated. This is repeated for each polygon, without the need to pre-sort the polygons as in other algorithms. Hence, objects appear on the screen in the order they are processed and not necessarily either front to back or back to front.

A further example of the image space category of algorithms is the scan line algorithm. The approach has the strategy of finding the visible polygons for each scan line of the image.

This simplifies the domain of the problem from polygons to lines. In order to reduce the number of polygons that have to be considered for each scan line, the polygons are sorted in one direction, (from the top downwards, for example). There are many variations of this algorithm, with different strategies in finding visible lines and how they propagate relevant information to neighbouring scan lines and capitalise on scan line and edge coherence.

Anti-aliasing

Many computer graphic images displayed on raster display devices exhibit disturbing image defects such as jagging of straight lines, distortion of very small or distant objects and the creation of inconsistencies in areas of complicated detail. These distortions are caused by improper sampling of the original image and are called aliasing artifacts. Techniques known as anti- aliasing, which have their roots in sampling theory, have been developed to reduce their influence (Crow 1977).

Shading

The next step towards the goal of realism is the shading of visible surfaces within the scene (Newell et al. 1972). The appearance of a surface is dependent upon the type of light source(s) illuminating the object, the condition of the intervening atmosphere, the surface properties including colour, reflectance and texture, the position and orientation of the surface relative to the light sources, other surfaces and the viewer. The objective of the shading stage is to evaluate the illumination of the surfaces within the scene from the viewer's position. The more oblique the surface is to the light rays, the less the illumination. This variation in surface illumination is a powerful cue to the three dimensional structure of an object.

There are two types of light sources apparent in the environment: ambient and direct. Ambient light is light reaching a surface from multiple reflections from other surfaces and the sky, and is normally approximated by a constant illumination on all surfaces regardless of their orientation. This simplified model produces the least realistic results since all surfaces of an object are shaded the same as experienced, for example, under heavily overcast cloud conditions. However, more complex modelling of the ambient or indirect component of lighting has produced enhanced realism using ray tracing techniques to model the contribution from specular inter-reflections and transmitted rays and radiosity techniques to account for complex diffuse inter-reflections.

Images are created by determining the visible surface at each screen pixel location, and then computing the intensity of light leaving a point in the direction of the viewpoint (eye). In simple shading models, the ambient illumination component is considered to be constant throughout the environment and only directly reflected light is considered, without considering inter-reflections from other surfaces.

Having modelled the intensity of colour at a point, this must now be expanded to encompass a surface. Most rendering algorithms require all surfaces to be decomposed and described as polygonal meshes. Since the normal of a polygon is constant over its area, the shade will also be uniform. Therefore, at polygonal boundaries there will be shade discontinuities that reveal the polygonal approximation of an object's structure. An example of this facetted shading is shown in Plate IV. Techniques have been developed to overcome this limitation. In Gouraud's approach (Gouraud 1971) the true surface normals at the vertices of each of the polygon are calculated. When the polygon is converted into pixels, the correct intensities are computed at each vertex and these values are used to linearly interpolate values across the polygonal surface. This approach eliminates the impression of underlying polygons. However, the approach is prone to distorting highlights and the production of Mach band interference effects.

An alternative, but computationally more expensive approach was developed by Phong (1975). This applies linear interpolation of the surface normals over the polygonal area and performs an explicit shading calculation at each of the pixels instead of interpolating the value. The result is normally superior to the Gouraud approach. Plate V is an example of the Gouraud approach and disguises the underlying polygonal surface structure.

Ray tracing

The approaches to shading so far described have used approximations to model the illumination environment. These ignore or excessively approximate illumination components resulting from light reflected from other surfaces. The model does not include the refracted component of light and becomes convoluted in treating complex multiple light sources. A more elegant, but computationally expensive solution is to use ray tracing techniques evolved by Appel, (1968), Whitted (1980) and Cook et al. (1984).

This approach, which is view dependent, involves tracing a ray of light from the viewpoint through a pixel and into the model where its interaction with objects is analysed (Figure 6.4). As the ray strikes the first object in its path, the ray is broken into three components: diffusely reflected light, specularly reflected light and transmitted (refracted) light. Similarly, a ray of light leaving the surface of an object is in general the sum of the three components. Therefore, each collision with an object generates three new rays that should ideally be traced. A diffuse reflection generates an infinite number of rays. Therefore, only rays from specular reflection and refraction are normally continued to be traced. However, Kajiya (1986) has included the modelling of complex diffuse interreflections in ray tracing, previously only modelled by Radiosity techniques. This collision and ray splitting procedure is continued until there is a cut-off point imposed based on the number of surfaces encountered or on a minimum energy level. This in effect grows a tree of rays from which the intensity at the pixel can be calculated.

This ray tracing procedure is repeated for rays passing through each pixel centre until the entire scene is generated. To avoid aliasing effects, multiple rays can be passed through a pixel and an adaptive process can be implemented to fire a varying number of rays per pixel, based on the complexity of the objects encountered by the rays.

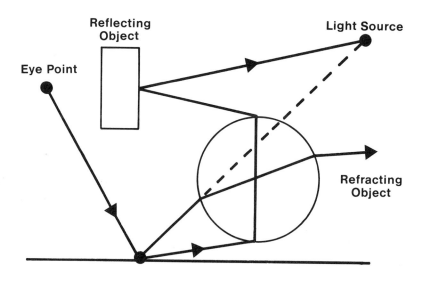

Figure 6.4. Principle of ray tracing.

Shadows

Shadows are an essential scene component in conveying reality in a computer graphics image. A scene that appears "flat" suddenly comes to life when shadows are included in the scene, allowing the comprehension of spatial relationships amongst objects. Shadows provide the viewer with one of the strongest visual cues needed to interpret the scene. They are not always present in a scene, such as an overcast day simulation, but are often essential.

The intricacies of the shadow algorithm used is related to the complexity of the model of the light sources. Computer graphic shadowing techniques can now handle light sources inside and outside the field of view, an infinite number of light sources, umbra and penumbra effects, a variety of wavelengths of light, as well as complex geometries of light sources including point, linear, area, polyhedra and sky light (Nishita and Nakamae, 1986). The fundamental approach is equivalent to the hidden surface algorithm approach. The shadow algorithm determines what surfaces can be "seen" from the light source. Those surfaces that are visible from both the viewpoint and the light source are not in shadow. Those that are visible from the viewpoint, but not from the light source are in shadow.

Surface texture detail

Natural landscape scenes are characterised by features with a wide variety of complex textures. Computer graphic visualisations of landscapes can only achieve an acceptable level of realism if they can simulate these intricate textures. The "flat" shading algorithms, described in the previous section, do not meet this requirement directly since they produce very smooth and uniform surfaces when applied to planar or bicubic surfaces (Plate V). Therefore, the shading approach must be supplemented by other techniques to either directly model or approximate the natural textures.

Atmospheric attenuation

Due to atmospheric moisture content, objects undergo an exponential decay of contrast with respect to distance from the viewpoint. The decay converges to the sky luminance at infinity. This reduction rate is dependent upon the season, weather conditions, level of air pollution and time. The result is a hazing effect.

The attenuation can be effectively modelled to simulate a variety of atmospheric conditions (Middleton, 1952). The intensity of the component primary colours decays exponentially by varying amounts, with blue being the most prominent at distance resulting in the blueness of distant objects. By introducing this atmospheric model into the rendering process, more realistic landscape scenes can be achieved and it introduces an important cue for depth interpretation.

Applications of digital terrain and landscape visualisation

The use of visualisation techniques for both military and civilian applications is an area of significant growth at present and some of the factors which account for this have been outlined previously. Generally these applications are either for small scale, regional visualisation or conversely for large scale, site specific projects. The former group are oriented primarily towards military applications whereas the latter tend to be more biased towards engineering, mining and landscape planning.

Small scale, regional digital terrain and landscape visualisation

For the majority of small scale regional applications the accuracy requirements of the height data are relatively low and may therefore be acquired from existing map sources. Because of the need for wide coverage it is also common practice for such applications to use one or more of the available regional terrain databases.

Cartographic applications

Quality control

One of the main difficulties associated with the production of large terrain databases is that of identifying gross errors at the data acquisition phase. Visualisation techniques have been developed to assist with this activity and a number of alternative strategies are discussed by Faintich (1984). For low resolution analyses of datasets the use of pseudo-stereo pairs of shaded relief images is suggested as having potential for the identification of spike points. For higher resolution anomaly analyses the use of convolution (edge enhancement) filters and fourier analysis techniques are highlighted. At a simpler level many organisations validate their terrain data by regenerating contours from the terrain database and comparing this with the original contour sheet.

Relief shading

Variations in elevations are normally represented symbolically by contours on topographic maps. Occasionally hill or relief shading is added manually by the cartographer in order to enhance the topography. Several authors have considered the use of regional DTMs for automating the relief shading process and Schachter (1980) and Horn (1982) provide some research results.

More recently Scholz et al. (1987) have discussed the procedure adopted to automate the production of relief shading on 1:250,000 scale aeronautical charts (Plate VI). The algorithm used simulates the effect of illumination by computing the slope and aspect for each point on the surface The authors estimate cost savings of up to 75% compared with the manual techniques previously used.

Remote sensing applications

DTM data have several possible applications in remote sensing particularly when regional data are available. For example they have been used to improve the geometric quality of satellite imagery, or to improve the accuracy of the supervised classification techniques which are currently being used. In the latter case the DTM is used to remove the variations in radiance caused by varying surface gradients since these effects are additional to the radiance caused by changes in land cover. In the field of visualisation, the main use of remotely sensed imagery is in the production of realistic perspective views of the terrain using DTM data merged with aerial or satellite imagery.

Military applications

DTMs are used extensively by the armed services for a number of applications ranging from general surveillance and intelligence gathering to radar simulation, battlefield planning and simulation and missile guidance. Although the techniques used by the military user do not differ substantially, their needs in this field differ significantly from those of the civilian users. Firstly, there is a need for more accurate and up to date terrain data. Secondly, there is a requirement for fast and effective methods of displaying the analyses of the terrain database. Finally it is essential to have efficient methods of communicating and distributing these results to users.

The development of portable field systems which can generate terrain information such as intervisibility analyses, perspective views, slope and aspect maps and assessments of cross country movement is a topic which has been widely reported in the USA. Tindall et al. (1981) describe some early work on the Field Exploitation of Elevation Data (FEED). More recently Sither, (1987) discusses the Terrain Analyst Work Station (TAWS) being developed by the US Army Engineer Topographic Laboratories (USAETL).

Large Scale, site specific terrain and landscape visualisation

Landscape planning - visual impact analysis

Growing public awareness of environmental issues has been recently strengthened by the European Economic Community's Directive on the "Assessment of the Effects of Certain Public and Private Projects on the Environment". This Directive will force proposed changes to the landscape to be publicly assessed for environmental impact. A component of this environmental audit is a statement on the visual intrusion of proposed landscape changes. Consequently, projects such as road construction, transmission line routing and open cast mining as well as more dynamic phenomena such as forestry will need to be visually judged.

Traditionally, landscape visualisation techniques have involved the building of physical models or the creation of artist's impressions. However, these are time consuming to create, and are inherently inaccurate and inflexible once created. In order to more accurately quantify the level of visual intrusion, computer graphic modelling and visualisation techniques are increasingly being used in the planning and design of landscape projects. These new approaches allow more accurate visualisations and more analytical assessments of visual intrusion to be determined. Due to the flexibility of the approach, many more proposed designs can be evaluated, resulting in a more refined design solution.

Turnbull et al. (1987) pioneered the development of a computer aided visual impact analysis system (CAVIA) that has been used, for example, to provide evidence at public inquiries related to electricity transmission line routing through environmentally sensitive landscapes (Turnbull et al. 1986). Projects are typically performed at the sub-regional level with areas up to 40 x 40 km being analysed. The approach uses DTMs, landscape features and proposed design objects to produce an estimate of the visual intrusion. This visual intrusion toolkit includes intervisibility analysis to produce levels of visual impact, dead ground analysis, identification of the portions of the landscape forming a backcloth for the design object, situations where the design object appears above the landscape horizon and the identification of optimal locations for vegetation screen placement.

Road/traffic engineering

Visualisation has found a number of interesting applications in the field of road design. Many road engineering design systems are now offering visualisation capabilities. These form an integral part of the design process and allow the design to be subjectively assessed and refined for safety and visual intrusion in the context of its environment.

The Transport and Road Research Laboratory of the UK have developed (Cobb, 1985) a system to model and visualise road designs. There are various applications of the system.

Urban design

In recent years urban renewal has become an activity increasingly exposed to and controlled by public and royal opinion. Architects, in an attempt to alleviate public fears of a continuation of the "Kleenex Box" era, have turned to computer generated images to convince the public of the merits of their proposed building designs. Computer generated visualisations have become a fashionable marketing tool.

Although the architectural industry was one of the first application areas where computer aided design (CAD) techniques were applied, it is only recently that tools for creating high quality visualisations of the resulting building designs have been made available. This capability is a natural extension of the CAD process and many CAD system vendors are now supplying this capability as an integral part of their system or as an interface to foreign visualisation packages.

Since national large scale map series of urban areas normally lack the essential height information of the buildings, users are either being forced into the expense of an original survey or being inhibited in the use of these techniques. In an attempt to encourage the use of these techniques, the City of Glasgow in Scotland sponsored the creation of a digital model of the downtown core of the city (Herbert, 1987). This was performed by the University of Strathclyde and the model included the terrain, all major buildings, streets and rivers. Potential users are now being encouraged to use the "city database" to site communications equipment, derive tourist information as well as create visualisations for planning purposes.

Conclusions

Computer generated visualisations of digital terrain and landscape scenes are now widely accepted in many application areas as efficient technical analysis, design and marketing tools. Visualisation techniques have released the world from its traditional two dimensional approaches to display and, in so doing, have highlighted the three dimensional deficiencies in our sources of data in terms of availability and accuracy.

Many large scale maps have minimal terrain elevation data and the majority ignore the three dimensional aspects of surface features. Although the photogrammetric acquisition of DTM data is well established, to date much less attention has been directed towards the photogrammetric acquisition of three dimensional descriptions of significant objects on the terrain. In many cases the three dimensional co-ordinates of features such as buildings may be observed, although often the Z co-ordinate values are deemed superfluous and do not form part of the recorded dataset. Indeed the lack of data is currently inhibiting the wider application of many of these techniques.

In the GIS environment, visualisation techniques are recognised as an invaluable system component, aiding in the interpretation of spatially related phenomena and complex data analyses that takes the GIS a step beyond two dimensional polygonal overlay analyses. Many of the GIS vendors are including this capability in their systems to help cope in our understanding of the "fire hose" of data being produced by contemporary sources such as satellites.

Over the past decade, substantial advances in the realism of computer graphic generated visualisations have been achieved. The techniques of ray tracing and radiosity in modelling global illumination are currently forming the leading edge. Despite this progress, the visualisation of natural phenomena inherent in landscape scenes are still simplistic, forming "realistic abstractions" or "abstract realism". Our visual system is specialised and highly skilled at recognising them in all their subtle forms, making the objective even more elusive. At present, apart from flight simulation, the production of realistic scenes using the visualisation techniques discussed in this paper have been largely restricted to static applications (Figure 6.5). Similarly at present, dynamic visualisations are primarily restricted to wireframe 'abstractions'. For the future, however, the ability to perform dynamic visualisation with high levels of scene realism will become an increasingly attainable goal, although the explicit modelling of many natural phenomena will continue to be a significant problem. For example, many of the current landscape visualisation products use photographs, frame grabbed images or texturing techniques of scene components to avoid the computer graphic synthesis of natural phenomena.

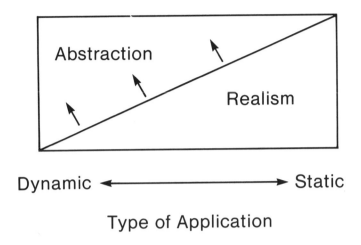

Figure 6.5. Future development of visualisation techniques

Some creators of the most realistic landscape images synthesised to date have used their artistic license by overly approximating the simulation of natural phenomena and 'faking' low interest scene components with stochastic or texture map processes. Although not creating a scientifically correct image, they have achieved surprisingly effective results using what could be termed artistic techniques. This emphasises the need for the scientific and artistic communities to work together towards this goal. Certainly more research is needed in perceptual comparison of real scenes and their synthetic image counterparts, with particular emphasis on the synthesis of natural colours and textures.

Inevitably, the successful approach will resort to true simulation of the phenomena based on the laws of physics. However, this is a computationally expensive option that in most cases is currently prohibitive, but will become feasible with the guaranteed increase in processing power. This is already happening with the recent release of parallel processing architectures, minisupercomputer based workstations, high bandwidth links between supercomputers and workstations and the availability of customised VLSI for specific applications. The present techniques of approximating or faking (Frenkel, 1988) will be displaced by progressive refinements of the simulation model. This approach has been endorsed by the McCormick et al. (1987) initiative on visualisation in scientific computing. Despite realism being a distant target, it acts as a convenient measure of our techniques and understanding and will continue to be relentlessly pursued to our continuing benefit.

Acknowledgements

This paper is based upon a more complete article entitled "Modelling For Digital Terrain and Landscape Visualisation", T.J.M. Kennie and R.A. McLaren, October 1988,

Photogrammetric Record, Vol. XII, No. 72, pages 711-746, and has been published with the full permission of the Photogrammetric Record.

The authors wish to express their sincere thanks to the many individuals who have provided information for inclusion in this review paper. In particular, they would also like to thank Cambridge Interactive Systems (CIS) and Laser Scan Laboratories who very kindly provided a number of colour photographs to be included in the paper.

References

Appel, A., 1968, Some techniques for shading machine renderings of solids. *AFIPS/ JSCC Proceedings*, **32**, 37–45.

Armstrong, A., 1976, A three dimensional simulation of slope forms. *Zeitschrift fur Geomorphologie, N.F. Supplementband*, **25**, 20–28.

Balce, A. E., 1987, Determination of optimum sampling interval in grid digital elevation models(DEM) data acquisition. *Photogrammetric Engineering and Remote Sensing*, **53**(3), 323-330.

Burrough, P., 1985, Fakes, facsimiles and facts: fractal models of geophysical phenomena. In *Science and Uncertainty*, edited by S. Nash, (Science Reviews Ltd.), pp. 151-169.

Butland, J., 1979, Surface drawing made simple. *Computer-Aided Design*, **11**(1), 19-22.

Clarke, K. C., 1987, Scale based simulation of topography. Presented paper to Auto-Carto 8, Baltimore, 9 pages.

Cobb, J., 1985, A new route for graphics. *Computer Graphics '85*, (Pinner, UK: Online Publications,) pp. 173-183.

Cook, R. L., Porter, T. and Carpenter, L., 1984, Distributed ray tracing. *ACM Computer Graphics*, **18**(3), 137-145.

Cook, R. L., Carpenter, L. and Catmull, E., 1987, The Reyes image rendering architecture. *ACM Computer Graphics*, **21**(4), 95-102.

Craig, R. G., 1980, A computer program for the simulation of landform erosion. *Computers and Geosciences*, **6**, 111-142.

CROW, F. C., 1977, The aliasing problem in computer generated shaded images. *Communications of the ACM*, **20**(11), 799-805.

Dubayah, R. O. and Dozier, J., 1986, Orthographic terrain views using data derived from digital elevation models. *Photogrammetric Engineering and Remote Sensing*, **52**(4), 509-518.

Durrant, A., 1987, Intelligent terrain modelling-the CV Medusa GIS Approach. Presented paper to Short Course on Terrain Modelling in Surveying and Civil Engineering, University of Surrey and University of Glasgow. 10 pages.

European Economic Community, 1985, Council directive of the 27th June on the assessment of the effects of certain public and private projects on the environment, (85/337/EEC).

Faintich, M. B., 1984, State-of-the-art and future needs for development of digital terrain models. *International Archives of Photogrammetry and Remote Sensing*, **25**(3a), 180-196.

Fournier, A., Fussel, D. and Carpenter, L., 1982, Computer rendering of stochastic models. *Communications of the ACM*, **25**(6), 371-384.

Frederiksen, P., Jacobi, O. and Kubik, K., 1985, A review of current trends in terrain modelling. *I.T.C. Journal*, **1985-2**, 101-106.

Frenkel, K. A., 1988, The art and science of visualizing data. *Communications of the ACM*, **31**(2), 111-121.

Gelberg, L. M. and Stephenson, T. P., 1987, Supercomputing and Graphics in the earth and planetary sciences. *IEEE Computer Graphics and Applications*, **7**(7), 26-33.

Gouraud, H., 1971, Continuous shading of curved surfaces. *IEEE Transactions on Computers*, **C-20**(6), 623-629.

Gregory, R. L., 1977, *Eye and brain*, 3rd edn., (London: Weidenfeld and Nicolson), 256 p.

Griffin, M. W., 1987, A rapid method for simulating three dimensional fluvial terrain. *Earth Surface Processes and Landforms*, **12**, 31-38.

Herbert, M., 1987, A walk on the Clydeside. *CADCAM International*, **6**(4), 41-42.

Horn, B. K. P., 1982, Hill shading and the reflectance map. *GeoProcessing*, **2**(1), 65-144.

Horn, B. K. P. and Bachman, B. L., 1978, Using synthetic images to register real images with surface models. *Communications of the ACM*, **21**(11), 914-924.

Kajiya, J. T., 1986, The rendering equation. *ACM Computer Graphics*, **20**(4), 143-150.

Mandelbrot, B. B., 1975, Stochastic models for the earths relief, the shape and fractal dimensions of coastlines, and the number area rule for islands. *Proceedings of the National Academy of Science U.S.A.*, **72**(10), 2825-2828.

Mandelbrot, B. B. and Van Ness, J. W., 1968, Fractional Brownian motions, fractional noises and applications. *SIAM Review*, **10**(4), 422-437.

McAulay, I. C., 1988, Visual descriptors. A design tool for visual impact analysis. Ph.D. Thesis., Plymouth Polytechnic. (Unpublished). 458 p.

McCormick, B. H., Defanti, T. A. and Brown, M. D., 1987, Visualisation in Scientific Computing - A Synopsis, *IEEE Computer Graphics and Applications*, **7**(7), 61-70.

McCullagh, M. J., 1988, Terrain and surface modelling systems: theory and practice. *Photogrammetric Record*, **12**(72), 747-779.

Middleton, W. E. K., 1952, *Vision through the atmosphere.* (Toronto: University of Toronto Press). 250 p.

Muller, J.-P. and Saksono, T., 1986. An evaluation of the potential role of fractals for digital elevation model generation from spaceborne imagery. *International Archives of Photogrammetry and Remote Sensing*, **26**(4), 63.

Newell, M. E., Newell, R. G. and Sancha, T. L., 1972, A new approach to the shaded picture problem. *Proceedings of the ACM National Conference, Boston*, pp. 443-452.

Newman, W. M. and Sproull, R. F., 1979, *Principles of Interactive Computer Graphics*, 2nd edn., (McGraw Hill International Book Company).

Nishita, T. and Nakamae, E., 1986, Continuous tone representation of three-dimensional objects illuminated by sky light. *ACM Computer Graphics*, **20**(4), 125-132.

Petrie, G. and Kennie, T. J. M., 1989, Digital terrain modelling. In *Engineering surveying technology*. Edited by T. J. M. Kennie and G. Petrie, (Blackie Publishing Co.) (in press).

Phong, B. T., 1975, Illumination for computer generated pictures. *Communications of the ACM*, **18**(6), 311-317.

Roy, A. G., Gravel, G. and Gauthier, C., 1987, Measuring the dimension of surfaces: a review and appraisal of different methods. Presented paper to Auto-Carto 8, Baltimore, pp. 68-77.

Schachter, B., 1980, Computer generation of shaded relief maps. In *Map Data Processing*, edited by Freeman, H. and Pieroni, G. G. (New York: Academic Press), pp. 355-368.

Scholz, D. K., Doescher, S. W. and Hoover, R. A., 1987, Automated generation of shaded relief in aeronautical charts. Presented paper to ASPRS Annual Convention, Vol. 4, 14 p.

Schwartz, M. W., Cowan, W. B. and Beatty, J. C., 1987, An experimental comparison of RGB, YIQ, LAB, HSV, and opponent colour models. *ACM Transaction on Graphics*, **6**(2), 123-158.

Sechrest, F. S. and Greenberg, D. P., A visible polygon reconstruction algorithm. *ACM Computer Graphics*, **15**(3), 17-27.

Sither, M. A., 1987, Terrain analyst work station (TAWS) demonstrations. Presented paper to ASPRS Annual Meeting, Baltimore, 9 p.

Snyder, J. M. and Barr, A. H., 1987, Ray tracing complex models containing surface tessellations. *ACM Computer Graphics*, **21**(4), 119-128.

Tempfli, K., 1980, Spectral analysis of terrain relief for the accuracy estimation of digital terrain models. *I.T.C. Journal*, **1980-3**, 478-510.

Tempfli, K., 1986, Composite/ progressive sampling-a program package for computer supported collection of DTM data. *Papers at ACSM-ASP Meeting, Washington, D.C.*, **4**, pp.202-209.

Tindall, T. O., Rosenthal, R. L. and Jones, C., 1981, Three dimensional terrain graphics for the battlefield. *41st ACSM-ASP Convention*, Washington, pp. 538-544.

Turnbull, W. M., Maver, T. W. and Gourlay, I., 1986, Visual impact analysis: a case study of a computer based system. *Proceedings of Auto Carto, London*, *1*, pp.197-206.

Turnbull, W. M., McAulay, I. C. and McLaren, R. A., 1987, The role of terrain modelling in computer aided landscape design. Presented paper to Short Course on Terrain

Modelling in Surveying and Civil Engineering, University of Surrey and University of Glasgow. 19 p.

Voss, R., 1985, Random fractal forgeries: from mountains to music. In *Science and Uncertainty*, edited by Nash, S. (Science Reviews Ltd.), pp. 69-86.

Whitted, T., 1980, An improved illumination model for shaded display. *Communications of the ACM*, **23**(6), 343-349.

Chapter 7

Computer-assisted cartographical 3D imaging techniques

Menno J. Kraak

Introduction

In mapping phenomena related to the earth, the three-dimensional real world has to be projected onto the paper's plane surface, for long the most important carrier of cartographic information. In some maps the cartographer was trying to represent the three-dimensional world as closely as possible, which was a difficult and laborious undertaking. Since the sixties, however, the importance of the computer as a cartographic tool has increased. Its use made some laborious manual techniques superfluous.

The application of computer graphics techniques introduced new aspects to cartography. It also made the practice of cartography change considerably, but the basic principles remained relatively unchanged (Morrison 1986). In disciplines such as molecular biology, architecture and engineering the application of three-dimensional graphics seems to contribute especially to the evolution of these disciplines.

This a trend which is also apparent in cartography. The following observations confirm this (Kraak 1988):

1. Computer technological developments–looking at developments in computer technology, it can be seen that computer systems are becoming smaller while their capacity increases. In addition the price of equipment is falling. A similar trend is seen when looking at the peripherals. These developments make the technology become more generally available and all disciplines can benefit from it. For cartography it may facilitate sophisticated map presentation.
2. Developments in computer graphics–since many computer graphics applications involve the display of three-dimensional objects and scenes, techniques were developed to display them on two-dimensional screens (Newman & Sproull 1981). These techniques deal with questions such as how depth, the third dimension can be displayed on a screen and how the three- dimensional world should be modelled in the computer so the images can be generated.
3. Developments in computer-assisted cartography–the introduction of the computer as a cartographic tool made cartographers change their approach to the discipline. Looking at today's cartography several trends can be distinguished.
   a) automation of the mapping and charting process;
   b) the production and use of thematic mapping software;
   c) the interest in the cartographic component of Geographical Information Systems;
   d) the interest in the development of cartographic expert systems;

e) an interest in 'new' cartographic products.

Trends c) and d) demand special attention. Geographical Information Systems are tools for an effective utilization of large volumes of spatial data. They represent the intersection of disciplines such as surveying, remote sensing, geography, geology and cartography. For planners, market researchers and policy makers it is possible to approach their problems in an integrated fashion. Due to the character of the problems to be solved there is a need for sophistication in data analysis, manipulation and presentation. The system's success heavily depends on its cartographic component, which presents its results. Here the third dimension plays an increasing role of importance as can be seen in other papers in this book.

The computer creates opportunities for the cartographer to work on map types which are, without the use of the computer difficult or laborious to produce (Taylor, 1984). This is certainly valid when the third dimension is involved. A greater interest in map types such as prism maps and digital terrain models can now be seen.

The combination of these observations instigated the research project described here: computer-assisted cartographical three-dimensional imaging techniques. Its objective is to see what characterizes three– dimensional maps and determine whether indeed three–dimensional maps produced by computer-assisted cartography give the map user a better understanding of the mapped phenomena. To be able to test this it is necessary to know what role the three-dimensional map plays in the cartographic communication process.

The next few sections deal with the cartographic discipline, 3D perception and 3D presentation techniques and theoretical aspects of three-dimensional cartography as well as 3D map production.

The map and 3D-presentation techniques

The cartographic approach

The project was approached by combining knowledge of cartographic theory, three-dimensional perception and computer graphics using new technological developments. In this approach cartography is seen as a combination of cartographic communication theories and the cartographic sign system, with emphasis on cartographic information analysis, the syntax of the sign system and 3D perception.

Figure 7.1 presents a simplified outline of this view on cartography. Three concentric circles can be located in the scheme, each describing several facets of the view. The inner circle indicates the main cartographic activities: map design, map production and map use. The location of the respective circle sectors is related to the middle circle, which portrays a simplified model of the cartographic communication process.

In the figure it can also be seen that the use of the computer as a cartographic tool decreases during the cartographic communication process. In the scheme's left half the sign system can be found. The upper left sector presents the process of cartographic information analysis. This is a procedure which lets the cartographer determine the character of the information to be mapped. It also provides links with the sign system's syntax.

The lower left sector of the outer circle contains the syntax of the sign system. The sign system or semiology contains rules to organize the signs in a graphic image. The syntax of this system indicates ordered relationships between signs or symbols. The study of the semantics, that is the relationship between symbols and their meaning, as well as the pragmatics, that is the relationship between the symbol and the map user, can be found in the right half of the scheme. The key-word here is perception.

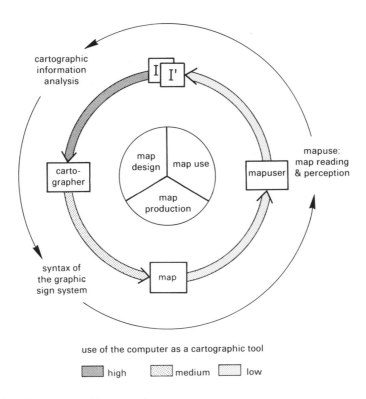

Figure 7.1 Outline of the cartographic approach.

The three-dimensional map

Together with written and spoken languages people use graphics to communicate ideas and concepts. As soon as a spatial component is included in the graphic image, cartography comes into focus. This spatial graphic image is called the map. No means of communication other than maps can give such a clear insight in this spatial component. Cartography deals with all aspects involved in the design, production and use of maps.

In the 'Nature of maps' (Robinson & Petchenik 1976) a map is defined as 'a graphic representation of the milieu'. This definition covers all graphical representations of spatial relations, however landscape drawings and paintings should be excluded. It is this definition of the map that is used here, and it is not affected by the computer's impact on the map. This is in spite of the introduction of a new state of display of the map, the temporary map. This new map exists next to the permanent and the virtual map.

- The *permanent map* is a map in a form familiar to most people. It is the traditional map presented mainly on paper, but also the globe and relief model belong to this category.
- The *virtual map* is the map in the mind. Two types can be distinguished. The first is a unique personal map image. An example is the mental map, shaped by one's knowledge and perception of the environment. The second is the map shaped in the mind when an expedient such as a stereoscope is used. This can be seen by more people at one time.
- The *temporary map* is the map which can be seen on a screen. It is produced by computer-assisted cartography and originates from a spatial database. In this map blinking and moving symbols can be introduced.

Returning to the definition of a map a question to be raised is 'Does it contain information on the dimensional aspects of spatial phenomena?'. It can be seen that they are not specifically mentioned. Three- dimensional maps are not included, but they are not excluded either. A fundamental question to raise now is 'what distinguishes them from two-dimensional maps and what are the characteristics of three-dimensional maps?'.

Maps are models of reality. They provide information on spatial phenomena existing in reality. They are constructed by a process of generalizing and structuring data from reality. Most spatial phenomena have a three-dimensional distribution, but by modelling them onto the two-dimensional map the third dimension is often not conveyed to the user. Sometimes it is omitted deliberately. For many applications this is not a problem since the map user is very well able to understand the model. For instance, when a map user visualizes the information retrieved from a city plan, a three-dimensional virtual image will be the result. This reduction from three to two-dimensions is often required for a pragmatic reason; most maps are presented on a flat medium. The absence of information on a spatial phenomena's third dimension qualifies the map as two-dimensional.

If the cartographer wants to preserve the third dimension, the construction of the map becomes more difficult. Even though the computer can be used to help in its construction, most permanent and temporary maps are still presented on a flat medium. Extra stimuli have to be added to help the map user perceive the map as three-dimensional. By some these maps are called two-and-a-half dimensional instead of three-dimensional, since the third dimension is not tangible.

Here a map, considered as a graphic representation of the milieu, is said to be three-dimensional when it contains stimuli which make the map user perceive its contents as three-dimensional.

3D-presentation techniques

To create three-dimensional maps several techniques are available. Figure 7.2 presents a classification of these techniques. Not all techniques mentioned are regularly applied to cartography and combinations of different techniques are possible. In the diagram the presentation techniques are related to the possible map states of display: the permanent, temporary and virtual map.

They are divided into two main categories. Those resulting in a real three-dimensional representation, e.g. the third dimension is tangible, and those resulting in suggestive representations in which the third dimension is non-tangible. The globe, the relief model and the tactual map can be found in the first category. The second category is further divided into sub-categories, depending on the number of images needed to create the three-dimensional map.

Examples of the first sub-category, the one-image representations, are maps presented on a two-dimensional medium, but with the necessary stimuli to let the map user perceive it as three-dimensional. Also images created by movement parallax, as well as mental maps belong to this category. The second sub-category of three-dimensional suggestive maps, the two-image representations, has to be looked at using a special device to perceive it as three-dimensional. The techniques mentioned in the last sub-category need several images to create a three-dimensional map.

The choice and usefulness of one of these presentation techniques depends on the interaction of three factors. These are:
• Human skills– these can have a physiological as well as a psychological character. For instance a small percentage of people are not able to see depth in a stereoscopic image, which

might limit the use of this technique. And people who suffer from colour blindness can have trouble with anaglyph maps;

• Purpose of the three-dimensional image– depending on the nature of the information to be transferred by the map and the character of the map's target group, the level of detail and realism to be included in the map image may vary. Some of the presentation techniques are more suitable for a very detailed image than others;

• Technical opportunities– the choice of a technique may be limited by pragmatic reasons, since not all cartographers will have all necessary equipment and materials available needed for a specific application.

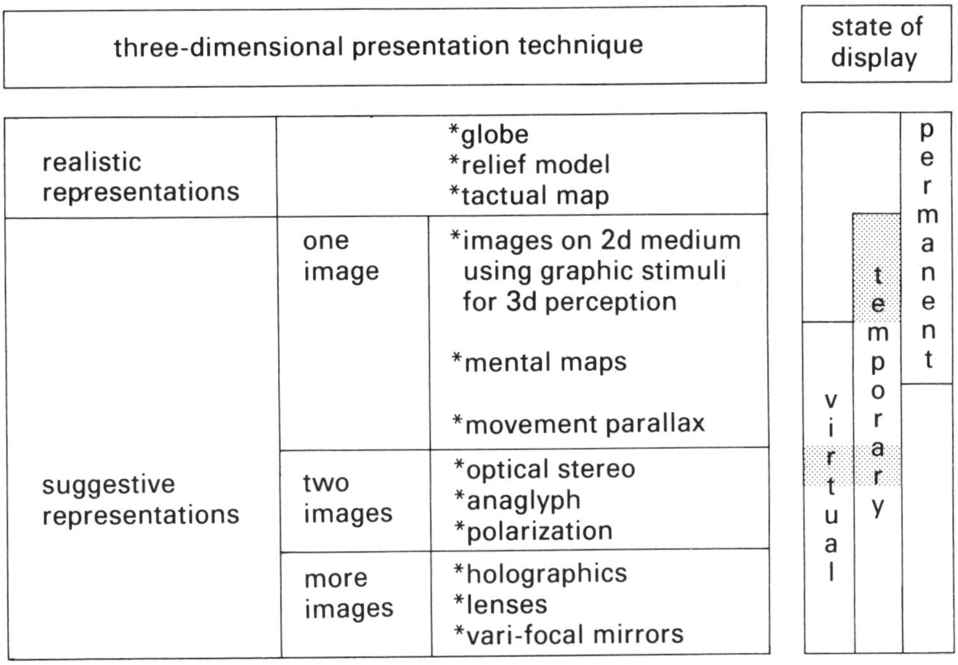

three-dimensional presentation technique			state of display		
realistic representations		*globe *relief model *tactual map	virtual	temporary	permanent
suggestive representations	one image	*images on 2d medium using graphic stimuli for 3d perception *mental maps *movement parallax	virtual	temporary	permanent
	two images	*optical stereo *anaglyph *polarization	virtual	temporary	permanent
	more images	*holographics *lenses *vari-focal mirrors	virtual	temporary	permanent

Figure 7.2 Classification system of three-dimensional presentation techniques in cartography. The patterned areas indicate the categories to which the three-dimensional maps studied belong.

The above definition of a three-dimensional map includes conventional maps such as layer-tint maps and shaded relief maps. However it is the non-orthogonal three-dimensional cartographic product which is interesting in this information age. The computer provides the cartographer with the opportunity to manipulate and experiment with its full three-dimensional map data set. These are also the products which Taylor (1984) had in mind when discussing the products of the 'new' cartography. These maps, those with a topographic as well as a thematic nature will be referred to as Spatial Map Images and were studied in this research project.

A Spatial Map Image can be described as a three-dimensional non– orthogonal representation of spatial phenomena. Examples are the digital terrain model, the prism map, the three-dimensional point symbol map and the three-dimensional urban map (Plates VII and VIII).

3D-perception

Vision

In designing maps the theory of the graphic sign system and the rules of legibility only provide information for the construction of two- dimensional maps. What has to be added to the map to let the map user perceive it as three-dimensional? The map was described as a model of reality above. Is it possible to obtain additional ingredients for a three-dimensional map from this reality?

Therefore its necessary to know how humans see and perceive the three- dimensional world around them. This results in knowledge of depth perception. Expedients which can be used to enhance the appearance of three-dimensional maps are also discussed.

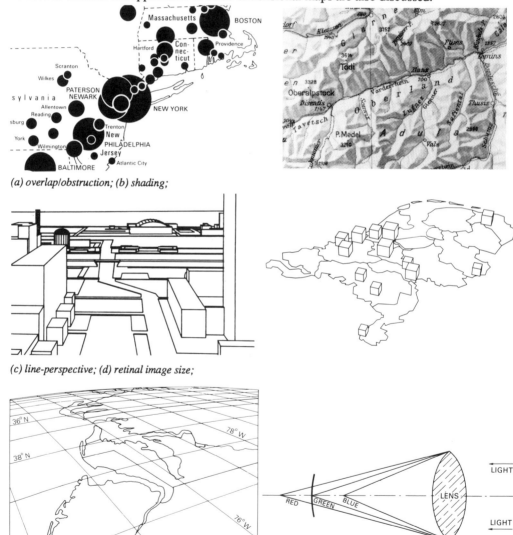

(a) overlap/obstruction; (b) shading;

(c) line-perspective; (d) retinal image size;

(e) texture gradient (f) Chromostereopsis
Figure 7.3 Psychological depth cues

Marr (1982) describes vision as an information processing task. Vision provides information on shapes and spaces as well as on spatial relations. The purpose of vision is to make a description of the shape and position of things from images; that is to obtain a representation of images.

All theories on vision acknowledge the existence of depth information. This depth information can be seen as a set of keys to allow a correct impression of depth. The depth cues can be divided into those related to physiological characteristics of the visual system, also called the physiological depth cues, and those related to the objects' structure and the way they result in images, also called psychological or pictorial depth cues. The second type of depth cues, such as shading, texture and perspective, can be used as stimuli in the map image to enhance the three-dimensional impression. The first category of depth cues can be used by applying expedients such as a stereoscope. An extensive explication of these depth cues can be found in Okoshi (1976), Gibson (1979), Hodges et al. (1985).

Psychological depth cues

Obstruction/ overlap or interposition

An object will be perceived as further away, but still complete, from the observer when it is partly hidden by another object. From this situation relative distances can be derived.

Shading

In combination with a known location of a light source shadow and shading provide information on the position (volume and distance) of objects. Humans are used to the fact the light source is located 'above' objects seen.

Perspective

Perspective is probably the best known pictorial depth cue. Because of its impact on the image's geometric relationships it influences its visual characteristics. For maps it should be noted that next to the actual perspective construction rules, the elevation angle (the angle formed by the horizon and an imaginary line of sight) and the slope (the angular relationship between the topography or the statistical surface and the horizon) play important roles in its three-dimensional perception. Several more specific depth cues can be derived from perspective:

- Line perspective– is created by the use of one or more vanishing points in the image. Lines which are parallel in reality seem to converge to one point;
- Retinal image size or size perspective implies that a larger distance to an object results in a smaller object in the image, and conversely;
- Texture gradient as a depth cue means that the texture of objects appears coarse in the foreground and grows finer as the objects recedes in the background;
- Aerial perspective is created by the fact that objects at a great distance tend to become hazy and bluish because of atmospheric scattering of the shorter wavelengths;
- Detail perspective implies that details become less visible at a greater distance, because of the limitations of the human visual system.

Colour

Colour is also one of the depth cues. Here it is important to know that the lens of the eye is not corrected for differences in wavelengths. This makes the eye accommodate constantly when looking at a multi- coloured image. The need to accommodate becomes strongest when looking at colours from both ends of the spectrum, such as red and blue. When red and blue are viewed from the same physical distance, they are not perceived as such, since red seems closer then blue. This effect is called chromostereopsis (Murch 1984 and 1985).

In many three-dimensional images the psychological depth cues are used in combination with each other. These combinations do not necessarily result in a better three-dimensional impression, and can result in an optical illusion. In chapters on three-dimensional graphics in general textbooks on computer graphics (Newman and Sproull, 1981 and Foley and Van Dam, 1984) techniques are given to apply these depth cues. To effectuate visual realism in the image perspective projection, intensity depth cueing, clipping, hidden line and hidden surface removal, as well as shading, can be used. In the next section the psychological depth cues will be compared with the graphical variables (Size, Value, Texture, Colour, Orientation and Shape) used in cartography.

Physiological depth cues

The physiological depth cues are accommodation, convergence, retinal disparity and movement parallax.

Accommodation

Accommodation is a change in thickness of the eye's crystalline lens caused by increasing or decreasing tension on the lens by the ciliary muscle in order to focus on an object.

Convergence

Convergence is a depth cue which originates from the angle both eye's viewing axes make with each other when both eyes are focused on one point. For an object close to the viewer this results in a wide angle. This angle of convergence is a cue to determine distance.

It has been found that there exists an interaction between accommodation and convergence. However, many workers (Rock, 1984 and Overbeeke and Stratmann, 1988) question the value of both accommodation and convergence as depth cues.

Retinal disparity

Retinal disparity is created by observing one specific scene with both eyes which results in two different images, one on the retina of each eye. This is due to the fact that both eyes observe the scene from a slightly different position; as human eyes are about 6.5 cm apart on average. The perception of depth which originates from observing two unequal images with two eyes is called stereopsis.

Monocular movement parallax provides depth information by movement of the observer linked to displacement in the image.

In three-dimensional images the physiological and psychological depth cues can be combined. The result of these combinations can weaken or strengthen the depth impression, depending on the type of combination. It can be that the use of a depth cue, such as overlap/ obstruction nullifies the effects of retinal disparity (Rock 1984).

Visual expedients

This paragraph will give an overview of possible expedients which can be used to enhance the three-dimensional perception of an image. The overview is related to the three-dimensional presentation techniques discussed in the scheme in Figure 7.2, and is not intended to be complete. Contemporary literature provides an extensive overview of these techniques and all its derived products (Okoshi, 1976, Hodgess et al, 1985 and Overbeeke and Stratmann, 1988). Expedients which are related to two-image representations are emphasized here. In related disciplines such as photogrammetry the experience of working with such expedients is extensive, and can be transferred easily to cartography.

3D information analysis and the graphic sign system

The general approach to cartographic information analysis for three- dimensional maps is not different from the two-dimensional case. Knowing the purpose of the map, and having defined the information and user requirements, the cartographer has to analyse the information which is to be mapped. An extensive description of this process is given by Kraak (1988). However, it is interesting to see how the geographical component (GEO) in a three-dimensional map differs from its two-dimensional counterpart. In a two-dimensional map the geographical component (further called GEO2) is seen as a single component using the plane's two dimensions (see Figure 7.4(a)). When there are more components to map, graphical variables are introduced (Figure 7.4(b)).

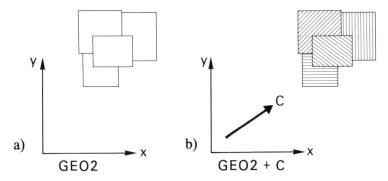

Figure 7.4 The geographical component in a two-dimensional map: (a) the geographical component using the two-dimensions of the plane (GEO2 = X+Y); (b) the geographical component and one other component (GEO2 + C).

How can a geographical component in a three-dimensional map (GEO3) be described? In addition to the two dimensions of the plane, X– and Y–, a Z–coordinate is introduced (Figure 7.5(a)). In GEO3 the Z-coordinate can represent a statistical value (examples are the prism map and the three- dimensional point symbol map), or it can derived from 'tangible' data from reality (examples are the digital terrain model and the three- dimensional urban map). It is also possible to introduce more than one component into the three-dimensional map: this will introduce graphical variables (Figure 7.5(b)).

What is the relevance of these questions? If the GEO3 equals GEO2 the information contents, in terms of results of the cartographic information analysis, of a three-dimensional map is larger then that of a two- dimensional map. To understand this the Map-To-See has to be introduced, and its relation to the temporary map, as well as to the Spatial Map Image has to be presented.

The concept of a Map-To-See or Image was first introduced by Bertin (1967), and is a clear graphic representation which can be comprehended in a short moment. Such an efficient graphic construction should provide the map user with answers to all possible questions related to the map, whatever their type or complexity, in a single moment of perception.

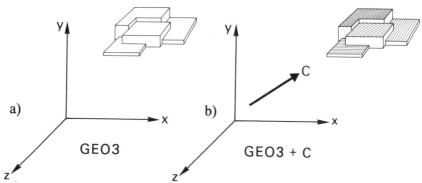

Figure 7.5 The geographical component in a three-dimensional map: (a) the geographical component and the two-dimensional plane (GEO3= GEO2+Z); (b) the geographical component and one other component (GEO3 + C).

A Map-To-See is constructed using the homogeneous plane (GEO2) and one visually ordered graphical variable. In the cartographic communication process the Maps-To-See are the most efficient means to communicate a message.

The importance of knowing the characteristics of GEO3, and it functions in the Map-To-See concept, will be understood when the strong relation between three-dimensional maps and computer-assisted cartography, more specifically the temporary map, is considered. Especially the use of the Maps-To-See, as temporary maps, will increase in this information age (Bertin and Muller in Taylor 1985). The characteristics of most screens which can display the temporary map do not allow for complex maps. These complex maps are called Maps-To-Read.

New information systems presented to the map user include TELIDON, VIDEOTEX, but also several navigation systems. They all use relatively small screens, often with a low to medium resolution. Since the maps presented by these systems are computer generated, there is an opportunity to use three-dimensional maps. However, the actual use of this type of map should depend on the application.

The graphic sign system and the third dimension.

The impact of the third dimension on cartographic theory is most significant with the application of rules of the sign system in three- dimensional maps. Most concepts in Figure 7.1 have, according to Bertin (1983), only a meaning in two-dimensional graphics. This does not imply that the rules are invalid, but research has not been undertaken on the application of the concepts to three-dimensional graphics.

The basic concept of the sign system is the two-dimensional plane. This is changed to a three-dimensional space in view of this research approach. There is no doubt of this change when virtual maps are involved. However, for temporary maps, most often presented on flat screens, this change is not a necessity. For these maps it is important to understand the qualities of the geographical component GEO3. It is this geographical component which determines the magnitude of the change. The change to the three-dimensional space is unquestionable when not only the map design, but also the database from which the maps are retrieved, is involved.

Working in three-dimensional space, a fourth basic graphical element has to be added to those in use in two-dimensional graphics. In addition to the point, line and area symbols the volume symbol is introduced, as is explained by Hsu (1979) and Dent (1985).

GRAPHICAL VARIABLES to display spatial distributions	DEPTH CUES to display the third dimension
Size	Retinal image size
Value	Shading
Texture	Texture
Colour	Colour
Orientation	Line perspective
Shape	Perspective
-----	Area perspective
-----	Detail perspective
Visual hierarchy	Overlapping / Obstruction

Figure 7.6 The possible relation between the graphical variables and psychological depth cues.

A question to be answered is whether the perceptual qualities of the two-dimensional graphical variables are still valid in three-dimensional maps, since there is a strong link between them and the psychological or pictorial depth cues discussed above, and as can be seen in Figure 7.6. Referring to the scheme in this figure it can be seen what possible relation between each graphical variable and a similar psychological depth cue exists. Both are used because of their perceptual properties, the first to display spatial distributions and the second to create a three-dimensional impression.

How does this combination function in a three-dimensional map? Given the combination is possible, do they strengthen or neutralize each other? Let's look at an example.

In cartography the perceptual properties of value are used to express information with an ordered or selective level of organization. The best known application of value in cartography is the choropleth map. In this map differences in the intensities of a phenomena are displayed by differences in value. Figure 7.7(a) presents an example.

Variations in value are also used in three-dimensional images to create an impression of depth. However the terminology is different. Value is called shading, since its effect is obtained by the use of a light source which lights the object and results in shadow and shading.

Questions which arise are: 'What is the effect of the combined use of the graphical variable value and the depth cue shading?'; 'How does the use of an expedient such as a stereoscope influence the map reading task?'; 'Is the map user distracted from the 'choropleth data' by the shading of the Spatial Map Image?' Figure 7.7(c) shows a map where the perceptual properties of value as graphical variable and as depth cue are combined. Several Spatial Map Images are produced to find answers to the above questions. The graphical variable value is also combined with other psychological depth cues to see how value behaves in these situations.

Each of the above combinations has been tested. These combinations are approached from a cartographic point of view, using the graphical variables as a base. Test maps have been produced for the map user in which each of the graphical variables, in combination with the psychological depth cues, appear. The answers to the accompanying questions (the users' map reading tasks) should lead to conclusions on the advisability of using them together in a three-dimensional map.

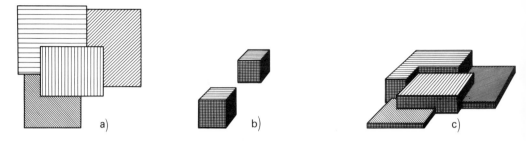

Figure 7.7 The perceptual properties of value: (a) value as a graphical variable; (b) shading as a depth cue; (c) value as a graphical variable, and shading as a depth cue, combined.

3D-map production

There are several software options for the three-dimensional map production. However, most programs do not includes all components necessary for a correct production of the maps. The basic principles of the three-dimensional map production system should be data collection and processing, the pre-display of the map, its manipulation and the final step in creating the map. The pre-display and manipulation of the map are an explicitly necessary when producing three-dimensional maps. Because there will always be dead ground in these maps information will be lost. To keep this loss to a minimum an ideal image position will have to be found, keeping the purpose of the map in mind. The final step in the map production process can be a hidden surface operation. An extensive description of the program developed can be found in Kraak (1987).

Before the most essential units of the program are treated, the software options, and several aspects of the computer graphics techniques applied will be explained.

Some existing cartographic software packages are able to produce three- dimensional maps, but they lack important principles presented above. The alternative would have been the use of CAD software, a package such as Autocad or CATIA. They have most of the functionality proposed here, and can be used to produce the Spatial Map Images. The reason they were not used is that they often have a restricted data structure, based on simple graphical primitives or mathematical descriptions.

In the near future software available might well be suitable for producing Spatial Map Images following the philosophy presented here. But at the moment a disadvantage of both the existing cartographic, as well as the CAD, software is that it is almost impossible to manipulate the final images, and to use them in the map user test. Therefore a program was specially developed for this study using a graphical standard, PHIGS, to be able to tailor the program to the needs of this study.

The program which displays the Spatial Map Images on a screen is written in FORTRAN using graPHIGS, IBM's version of PHIGS (Programmer's Hierarchical Interactive Graphical Standard) PHIGS is a detailed description of graphics functions, error conditions, and Fortran, C and ADA language bindings. It is intended to provide a common programming base for graphics hardware and application program developers, and is used for the actual drawing and manipulation of the Spatial Map Images. PHIGS routines function like those of GKS (Graphical Kernel Standard), and have at present the status of an ISO working draft (Plaehn, 1987 and Singleton, 1986). It is used to guarantee hardware independent software, and was preferred to GKS since it has special facilities for three-dimensional applications. The graphical primitives used are defined in three-dimensions (i.e. in X– Y– and Z–coordinates).

The use of a graphic standard has some important advantages. The program will be less hardware dependent, have a prolonged life span, and the application programmers can work more efficiently since they have to know only one standardized set of graphic concepts. Other positive features are the development of a set of concepts, the software evaluation possibilities that are contained in the standard, the simplification of the writing of device drivers, and the possible exchange of meta-files. Meta-files contain information on the graphic image only, and can be used to draw the image directly.

A disadvantage of the use of graphical standards can be that facilities such as shading and hidden surface calculations have to be produced by the programmer himself, while commercial packages have these functions often as standard options. The fact that the programmer/user has to write the whole program himself is by some seen as another disadvantage. These standards should therefore be considered as guidelines, and not as the definitive answer to all graphics software development and exchange problems, since there will always be applications which are too specific for the use of the standard.

Conclusions

A computer-assisted map user test was conducted in which map users are confronted with Spatial Map Images to find answers to assumptions made and questions raised when discussing the three-dimensional approach to cartography. Before starting the test attention was given in to the characteristics of the test map, the test environment, the subjects, the questions and the test procedure. The test and its results are discussed in detail in Kraak (1988).

The purposes of the test were threefold, as discussed in the above sections: first, to see how the geographical component functions in a Spatial Map Image; second, to see how the graphical variables can be used in a three-dimensional map in combination with the psychological depth cues; and third, to measure the effects of the use of a stereoscope, the expedient chosen.

During this study the results of the computer-assisted map user test were only used to increase the insight into specific three-dimensional maps, the Spatial Map Images. In the total design process of maps in general the results of such tests should be used directly to enhance the map design. This gives the cartographer the opportunity to increase the effectiveness of the cartographic product. However, this implies a strong relationship between the target group of the map and the subjects participating in the test. The use of computer-assisted map user tests increases the influence of the computer in the cartographic communication process, and might well lead to a further overlap of I and I' in Figure 7.1.

Referring to the geographical component and the test results it can be said that the Spatial Map Image as a Map-To-See is best used to present an overview of a relatively simple spatial phenomenon. The relationships between the 'extreme' objects or values in the map can also be clarified.

The test revealed that the geographical component in a Spatial Map Image (GEO3) is equal to the geographical component plus one component in a two-dimensional map (GEO2+C); see figures 7.3 and 7.4. Therefore the Z in GEO3 is equal to 'C' in GEO2+C. This leads to the conclusion that when a Spatial Map Image must function as a Map-To-See, only the geographical component should be used. Complex three-dimensional maps such as a digital terrain model with a draping of land-use data can be used, but in the cartographic communication process it will function as a Map-To-Read.

The different Spatial Map Images in the computer-assisted map user tests included each of the graphical variables, with or without the depth cues applied. However, the results of the tests do not reveal a specific pattern for all graphical variables together. Results here should be interpreted as indicative and further testing on this subject currently takes place.

The purpose of part of the computer-assisted map user test was to see whether the three-dimensional Spatial Map Images would be better understood when they really could be viewed by the subjects as three- dimensional (the virtual map). Therefore almost half of the test maps were presented to the subjects in stereo. A simple table stereoscope was modified and placed in front of the screen to view the stereo Spatial Map Images.

From the test results it can be learned that for the combined Spatial Map Images the response time is significantly shorter for the stereo maps compared with the mono maps. However the quality of the answers to the 'stereo-questions' does not differ significantly from the 'mono- questions'. Viewing a Spatial Map Image in stereo means a faster, but not necessarily a better, understanding of the map.

An improvement of the stereoscopic possibilities, for instance with a device with liquid crystal shutters, might shorten the response times to the 'stereo-questions' even more, and might even influence the quality of the answers given in a positive way, because only a primitive method was used in this study.

In discussing the three main topics of this research project the characteristics of the Spatial Map Images have become clear. Referring to cartography as a whole it can be said that the general cartographic theory can be applied, with a few exceptions as discussed above, to three-dimensional maps such as the Spatial Map Images. In the cartographic communication process these maps, provided they are kept relatively simple and are used as Maps-To-See, function at least as well as two-dimensional maps. For more complex maps further comparison between two- and three-dimensional maps will be necessary.

References

Bertin, J., 1983, Semiology of graphics, (Madison: The University of Wisconsin Press). Original french edition 'Semiologie graphique' published in 1967 by Mouton in Paris.

Dent, B. D., 1985, *Principles of thematic map design*, (Reading Mass.: Addison Wesley).

Foley, J. D. and Van Dam, A. 1984, *Fundamentals of interactive computer graphics*. (Reading: Addison-Wesley).

Gibson, J. J., 1979, *The ecological approach to visual perception*. (Boston: Houghton Mifflin).

Hodges, L. P., McAllister, D. F. and Robbins, W. E., 1985, *True three dimensional display technology and techniques for computer generated images*, (San Franscico: SIGGRAPH–coursenotes no.13).

Hsu, M. L., 1979, The cartographer's conceptual process and thematic symbolization. *American Cartographer*, **6**(2), 117-127.

Marr, D., 1982, *Vision*, (San Franscico: W.H.Freeman).

Morrison, J. L., 1986, Cartography: a milestone and its future. In *Proceedings Autocarto London*, *1*, pp.1-12.

Murch, G., 1984, The effective use of color. Physiological principles, Perceptual principles and Cognitive principles. *Techniques* **7**(4), 13-16, **8**(1), 4-9 and **8**(2), 25-31.

Kraak, M. J., 1987, Large scale thematic maps and solid modelling. In *Proceedings 12th UDMS, Blois*, pp 295-300.

Kraak, M. J., 1988, *Computer-assisted cartographical three-dimensional imaging techniques*, (Delft: Delft University Press).

Newman,W. M. & Sproull, R.F., 1981, *Principles of interactive-computer graphics*, (New York: McGrawHill Inc.).

Okoshi, T., 1976, *Three-dimensional imaging techniques*, (New York: Academic Press).

Robinson, A.H. and Petchenik, B. B., 1976, *The nature of maps*, (Chicago: University of Chicago Press).

Overbeeke, C. J. and Stratmann, M. H., 1988, *Space through movement*, (Delft: Faculty of Industrial Design).

Plaehn, M., 1987, PHIGS: Programmer's Hierarchical Interactive graphics standard. *Byte* **12**(13), 275-286.

Rock, I., 1984, *Perception*, (New York: W.H.Freeman).

Singleton, K., 1986, An implementation of the GKS-3D/ PHIGS viewing pipeline. In *Proceedings Eurographics '86*, edited by Requicha, A. A. G., (Amsterdam: North Holland)

Taylor, D. R. F., 1984, Computer assisted cartography, new communications technologies and cartographic design: the need for a 'new cartography'. In *Proceedings ICA Conference Perth, 1*,pp.456-467.

Taylor, D. R. F., 1985, The educational challenges of a new cartography. In *Education and training in contemporary cartography*, edited by D. R. F. Taylor, (Chichester: J.Wiley & Son), pp.3-25.

Chapter 8

The role of three-dimensional geographic information systems in subsurface characterization for hydrogeological applications

A. Keith Turner

Introduction

Most commercially available GIS products cannot handle true 3- dimensional data, although they can handle topographic data, usually as a digital elevation model (DEM), and display isometric views, contour maps and so on. Most DEM's use either gridded elevation matrices, or triangular meshes (TIN's) to allow for these terrain representations. In these cases, the elevation, or Z- coordinate, is treated as a dependent variable. Some systems allow the draping of another mapped feature, such as soils or land cover data, onto an isometric view of a topographic elevation surface, thereby creating an illusion of a 3-dimensional scene.

Common GIS applications involve the mapping of essentially 2- dimensional land surface phenomena such as land-use, forestry, or soils. Some geological applications can be accomplished by reducing the 3-dimensional representation to a quasi 2-dimensional one through the use of surfaces. These surfaces, which can represent bedding planes for example, can then be contoured or displayed as isometric views. However, in these cases, the elevation of the surface is not a truly independent variable, and so these systems are best defined as quasi- three-dimensional, or 2.5-dimensional systems. Many regional geological studies can operate in a 2.5-dimensional mode, because the geographical dimensions (X and Y) are several orders of magnitude larger than the depth dimension (Z).

The demands for detailed three-dimensional subsurface data are especially acute in such applications as petroleum reservoir characterization to support enhanced oil recovery (EOR), ground water contamination modeling at hazardous waste sites, and geotechnical site characterization for increasingly complex construction projects. All these applications have one thing in common; they all require increasingly quantitative and accurate rock property characterizations within the three-dimensional subsurface environment. Three-dimensional data are required because the depth dimension is in the same general range as the surface dimensions, and the true spatial relationships are important to the problem analysis.

Geological applications of GIS form but a small percent of the total GIS market; and GIS is much smaller, and distinct from, the even larger CAD/CAM graphics market. Furthermore, the inclusion of the third coordinate and the conversion from planar to solid geometry undoubtedly adds storage and computational overheads to the GIS software. Thus it is not surprising that, until very recently, developers and vendors of GIS software have not pursued the 3-dimensional GIS market. This situation is now beginning to change, due largely to the rising interest in 3-D graphics systems for many uses, and the development of affordable new hardware that can support the rapid generation of 3-D graphical displays.

The lack, to date, of affordable, fully-functional, 3-dimensional GIS products has had ramifications for many geological applications. This paper is restricted to analyzing the probable future role of 3-dimensional GIS technology to subsurface characterization for support of hydrogeological studies. Because appropriate 3-dimensional GIS systems are just now becoming available, there are no published examples of their use in these fields. Accordingly, this paper has two objectives:

1) to demonstrate the need for 3-dimensional GIS; and
2) to show conceptually how 3-dimensional GIS capabilities can be used, in combination with other modeling and analysis tools, to materially assist these studies.

The need for 3-dimensional GIS for subsurface characterization

Modern geological applications require increasingly quantitative and accurate rock property characterizations within the 3-dimensional subsurface environment. These applications differ from those faced by most other fields because of four major difficulties:

1) normally only very incomplete, and sometimes conflicting, information is available concerning the dimensions, geometries, and variabilities of the rock units, at all scales of interest, from the microscopic to the megascopic;
2) the natural subsurface environment is characterized by extremely complex spatial relationships;
3) economics prevent the sufficiently dense sampling required to resolve all uncertainities; and
4) the relationships between the rock property values and the volume of rock over which they are being averaged (the scale effect) are usually unknown. These difficulties are discussed further in the following sections.

Spatial visualization versus data management

True 3-dimensional GIS products can greatly aid the resolution of the above difficulties in two ways; by assisting the persons performing the analysis visualize the spatial relationships, and by providing data management services. To date, the visualization of these 3-dimensional features has been a major constraint. The ability to rapidly create and manipulate 3-D images can materially speed up the geoscientist's understanding of the subsurface environment. For example, colleagues of the author have been analyzing three-dimensional ground-water flow using accepted, publically available, models. Typical calculations take only a few hours on a powerful "386-class" personal computer. However the interpretation and visualization of the results from each model run, by contouring a series of 2-dimensional surfaces and slices using available software, takes a week or more.

By making true 3-dimensional visualization possible in real- time, or near real-time, 3-dimensional GIS can materially improve existing analytical capabilities. Recently announced commercial systems, such as the Dynamic Graphics Interactive Volume Modeling (IVM), have at least partially addressed these visualization needs (see Smith & Paradis, this volume).

Many types of geoscientific modeling require the extraction of information from large multi-parameter data sets, and the representation and modification of complex and uncertain geo-objects of interest. These applications require non-standard solutions to database access and 3-dimensional representation due to the range of data types, non-regular data distribution, the spatial access key, and the complex geometry of the subsurface domain.

Attempts to use commercial GIS methods and CAD/CAM technologies developed for other uses have been found to be only partially suitable for geoscientific applications. Accurate interactive 3-dimensional geoscientific modeling requires the development of a new spatial theory for complex objects, and the optimisation of the graphical data structure to allow for the varied attribute storage. The 3-dimensional solid modeling capabilities of CAD/CAM systems generally do not address the problem of representing complex objects known with differing degrees of confidence.

Much research work therefore remains to be done. The relative merits of the relational data model for geoscience database storage require evaluation, given the need for repeated spatial access to a wide variety of data types.

Ground-water flow modeling

Recent improvements in computer modeling capabilities have resulted in several economical and powerful 3-dimensional groundwater flow models. For example, Fogg et al (1983) used a three-dimensional integrated finite difference (IFD) mesh to study the ground water flow in the Wilcox-Carrizo aquifer system near the Oakwood salt dome in East Texas.

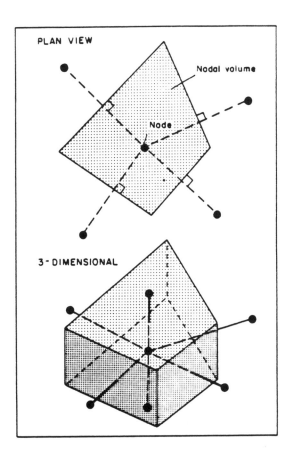

Figure 8.1 Example of a 3-dimensional integrated finite difference mesh (from Fogg et al. 1983)

Figure 8.1 compares the 3-D mesh to the more common 2-D case. Figures 8.2 and 8.3 show examples of the model mesh. These were developed using 2.5-dimensional general purpose graphics routines. Results of these studies are summarized by Fogg (1986). More recently, 3-dimensional coupled flow and solute transport finite element codes have been developed (Gupta et al. 1987).

These models have demonstrated that the spatial variability in hydraulic conductivity (K) is a critical factor affecting ground water flow and dispersion (Gelhar and Axness 1983). Other studies have suggested that the coefficient of dispersion used in the advection-dispersion equation is often incapable of characterizing the ability of the aquifer media to disperse solutes in ground water (Fogg and Kreitler 1981). More recent studies in the Texas Gulf Coast (Fogg 1986) have shown that when 3-dimensional models are used, the computed hydraulic head values are much less sensitive than are the flow velocities to changes in heterogeneity in K. Thus, the computed heads gave little indication of the presence of well-connected high-permeability zones, and even "well calibrated" models are likely to retain substantial errors in the prediction of dispersion and solute patterns.

Defining the "parameter crisis"

Recent experiences, such as those described in the previous section, have demonstrated that the latest generation of three- dimensional ground water and petroleum reservoir simulation models are capable of efficiently and accurately calculating the hydrodynamic flow characteristics of the fluids being evaluated, provided suitably accurate three-dimensional characteristics of the geological materials can be supplied. Some problems remain, however.

One major problem is the inability of the model users to rapidly display and visualize the results. The limitations of the 2.5-dimensional displays shown in Figures 8.2 and 8.3 are obvious. A second problem relates to the geological characterization of the modelled volume in 3-dimensions. The model results are sensitive to the selection of input parameters, and traditional model calibration methods may fail to identify problems. In fact these models have outstripped our ability to supply the necessary data using traditional methods. A "parameter crisis" faces those who wish to use such models.

The use of true 3-dimensional GIS products appears to be the best hope for solving both the spatial visualization and data management problems facing the users of these sophisticated flow models. This will require linkages between the 3-dimensional GIS programs and the models. However, 3-dimensional GIS techniques must also be linked to a variety of other analytical procedures in order to solve the subsurface characterization problem, and hence offer a solution to the "parameter crisis". The following sections discuss some possible analytical procedures which have been used or proposed for subsurface characterization.

Methods of subsurface characterization

It has long been recognized that the heterogeneities of an aquifer must be known in some detail in order to simulate or predict the transport of contaminants in a ground water system, or the depletion of petroleum resources within a reservoir. Identification of heterogeneities is often undertaken in association with field sampling, monitoring, or evaluation procedures. Because most petroleum reservoirs, and many ground water contamination problems, occur within sedimentary rock units, much attention has been directed toward their characterization and a number of approaches have been applied to the identification of heterogeneity within sedimentary rock environments.

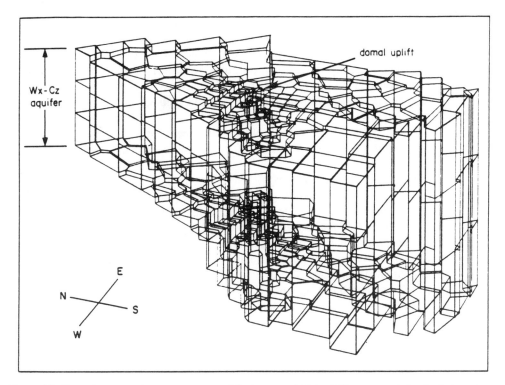

Figure 8.2 Three-dimensional perspective view of the Integrated Finite Difference mesh for the Oakwood salt dome, viewed from the southwest. (from Fogg et al. 1983)

An exact model of the subsurface is not possible. Simplified versions may be derived from geological and engineering data developed from exploration procedures, including borings, well tests, seismic explorations, and stratigraphic or sedimentological descriptions of depositional systems. Those model variables that are important for the characterization of the depositional system being studied may be represented by probability distributions. By these methods, descriptive depositional models may be converted into quantitative 3-dimensional subsurface representations. The ability of geologists and engineers to identify the depositional systems and to establish the parameters for all components is the most important constraint on how accurately the 3-dimensional representations will mimic reality.

Stochastic versus Deterministic Modeling

Subsurface characterization research has focussed on the inherent uncertainities associated with any definition of the subsurface. It is of course understood that the subsurface has potentially measurable properties and features at all scales of interest. The subsurface is intrinsically deterministic; it really exists! However, in order to determine these properties, every part of the region of interest would have to be excavated. The present subsurface conditions are the end product of many complex processes. For example, a sedimentary rock sequence may show the combined effects of erosion, transport, sedimentation, burial, compaction, and diagenesis of the sedimentary materials.

Faced with such complexities, it is not surprising that stochastic modeling approaches have been favored. A stochastic phenomena or process is one which is characterized by the property that its observation under a given set of circumstances does not always lead to the same outcome. So there is no deterministic regularity, but rather different outcomes occur in such a way that there is statistical regularity. Stochastic modeling or simulation in current geoscience, reservoir engineering, and hydrogeology usually refers to either:

1) generation of synthetic property fields in one, two, or three dimensions which possess a number of desirable features and hopefully parallel reality; or

2) studies of the effects of uncertainities in these generated synthetic probability fields on the responses of interest.

All deal with different aspects of the same problem: the effect of scale, parameter variability, and parameter uncertainity on the precision of the response estimates. There are many different approaches, including:

1) discrete grids versus continuum models;

2) random sampling with assumptions of frequency distributions;

3) estimates based on spatial co-variances (including kriging and co-kriging);

4) conditional simulation techniques which involve some type of deterministic modeling;

5) transitional probability (Markov) procedures; and

6) inverse approaches involving the back calculation of property variablity from variations in observed or estimated responses.

Stochastic methods have been used extensively for reservoir characterization methods in the petroleum industry (Augedal et al, 1986; Haldorsen et al, 1987).

Geologic process simulation models

Based partly on the concepts discussed above, many researchers have concluded that there are only two ways of increasing our knowledge of the subsurface:

(1) the "brute force" method of drilling large numbers of holes, perhaps supported by geophysical surveying techniques, or

(2) the use of geologic process simulation modeling to help characterize the most probable subsurface conditions based on limited exploration data.

The second method appears to hold the most promise for efficiency and applicability in many field situations.

Over the past two decades, a number of deterministic and/or stochastic geologic-process-simulation computer models have been developed for a number of sedimentary environments. These advancements have grown out of comparative studies of modern sedimentary environments, coupled with geophysical studies of ancient systems. For example, alluvial depositional systems formed by meandering rivers have been studied, first in a qualitative fashion (Allen 1965, 1974), and then in a series of progressively more comprehensive quantitative models (Leeder 1978; Allen 1978). These culminated in computer simulation models (Bridge and Leeder 1979; Bridge 1979). These studies developed the concept of the "interconnected-ness ratio", which measures the proportion of individual channel sand belt bodies which touch each other. More recently Fogg (1986) applied this ratio to the three dimensional modeling of the Wilcox aquifer system in Texas, using data from core logs and cores from over 100 boreholes in a 1300 square kilometer region.

The Bridge-Leeder model is applicable to meandering river systems. However parallel studies have been conducted for braided river systems (Cant and Walker, 1976; 1978; Miall, 1977; 1978). Similar studies have been conducted for a variety of other sedimentary environments, including glacial, eolian, and coastal systems. The reference literature is extensive. The interested reader is referred to Walker (1984) for more information.

Computer models have been developed for simulating several of these systems. Usually these models combine deterministic components, often using empirical formulae, with

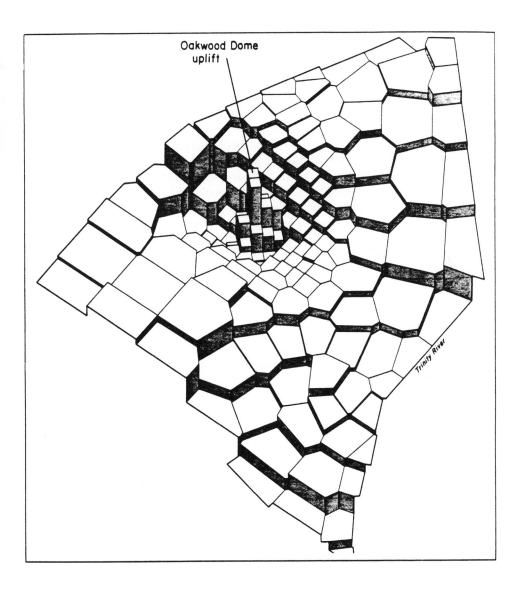

Figure 8.3 Three-dimensional perspective of the upper surface of the Integrated Finite Difference mesh for the Oakwood salt dome, viewed from the southeast. (from Fogg et al. 1983)

stochastic components in order to introduce a suitable level of complexity, or uncertainty, into the results. Measures of statistical or geometrical properties have demonstrated that these models replicate actual systems.

Use of these models can be considered a type of "Expert System" because they incorporate the expertise of many geologists in their formulation, and emulate the thought processes of an experienced geologist in developing a conceptualization of subsurface conditions from limited data. Many of the existing models are 2-dimensional, but can be readily expanded to 3-dimensions. Any such 3-dimensional versions will need to be linked to 3-dimensional GIS for both the visualization of their simulation results and for the data management functions required to convert these products into forms required by the ground-water models.

Markov process models

An alternative view has prompted some researchers to pursue another approach; Markov processes. "A Markov process is a process or feature, natural or artificial, that has an element of randomness or unpredictability, but in which a past event has an influence on a subsequent event. This aspect, in which a subsequent event 'remembers' a past event, is termed the Markov property" (Lin and Harbaugh, 1984). Many geological processes can be perceived as having this Markov property, which may range from very weak to highly dominant. Basic programs for Markov analysis were described by Krumbein (1967). Geological applications of Markov processes have been summarized by Harbaugh and Bonham-Carter (1970) and Davis (1986).

The use of Markov analyses have been almost entirely restricted to one-dimensional data sequences. A common application has been to stratigraphic analyses, in order to identify lithologic sequences and facies relationships (Carr et al. 1966; Miall 1973; Powers and Easterling 1982). However, Lin and Harbaugh (1984) have described extensions of Markov processes to two- and three- dimensional data.

In these cases, it is possible to analyze planar, or solid, patterns mathematically. It is also possible to generate a series of synthetic patterns which closely replicate the geometrical and statistical relationships contained within a supplied reference pattern. Such simulations can be used to generate a number of equally probable subsurface patterns. Each in turn can be subjected to hydrogeological flow modeling.

Once again, the visualization of the 3-dimensional results has been a major constraint. Lin and Harbaugh (1984) created stereoscopic images of each individual state (i.e. each facies) in order to try to visualize the results. This is cumbersome, and a major advance would be to link such simulations to a 3-dimensional GIS.

Hydrofacies definition

Many sandstone aquifers are actually multiple-aquifer systems composed of discontinuous sand bodies distributed in complex patterns within a matrix of lower-permeability silts and clays. The arrangement and interconnections among these facies stongly influence the hydraulic conductivity (K) and therefore the patterns and rates of groundwater flow and mass transport.

Hydrofacies are 3-dimensional sediment bodies displaying distinct hydraulic properties. In studies at the Hanford, Washington area, Gaylord and Poeter (1988) defined four hydrofacies within the shallow unconsolidated sedimentary aquifer based entirely on grain size and sorting characteristics. These hydrofacies differed from the more traditional lithofacies and lithostratigraphic units because they were defined solely on permeability and porosity considerations and ignored other geological formation aspects, such as depositional environmental factors.

Gaylord and Poeter were restricted to using only textural (grain size and sorting) factors in defining their hydrofacies because the drilling methods which had been used did not supply intact cores. They suggest that hydrofacies should be defined with consideration given also to the degree of cementation and the nature of sediment particle packing.

It seems that hydrofacies may be more readily converted to the input parameters required by many ground-water flow models than can lithofacies. This is because a single lithofacies may include a wide range of values for such hydrological factors as permeability or porosity, while a single hydrofacies should be more consistent in these factors. On the other hand, lithofacies may better define spatial geometries, since they are defined in terms of environments of deposition. Gaylord and Poeter (1988) suggest that both hydrofacies and lithofacies are required, and should be used, in a complementary fashion to fully characterize a subsurface sedimentary environment. Their studies were also constrained by a lack of suitable 3-dimensional visualization and data management capabilities.

Inverse plume analysis

Domenico and Robbins (1985) and Domenico (1987) developed methods to determine aquifer and contaminant source characteristics from the spatial distribution of contaminant concentration in a contaminant plume in homogenous aquifers. These methods are termed inverse plume analysis. Most field techniques, such as aquifer pumping tests, are designed to determine average or bulk parameter values for the volume of porous material being tested. Inverse plume analysis techniques use contaminant concentration data to determine the three orthogonal dispersivities, the center of mass of the contaminant plume, and the contaminant source strength and dimensions.

Inverse plume analysis has been performed with field data (Lavenue and Domenico 1986), and has ben verified using a 3-dimensional finite element code (Domenico 1987). The original technique was restricted to isotropic and homogeneous aquifers, but Belcher (1988) has extended the method to heterogeneous aquifers.

The role for 3-dimensional GIS

The preceeding discussions serve to define the role for 3- dimensional GIS in hydrogeological applications. Three-dimensional GIS methods have a central role to play because a properly configured GIS can support both the data management and the 3-dimensional visualization tasks which until now have been lacking. However, 3-dimensional GIS cannot, by itself, entirely solve the hydrogeological analysis problems.

The process of hydrogeological analysis can be considered in terms of four fundamental modules:
1) subsurface characterization;
2) 3-dimensional GIS;
3) statistical evaluation and sensitivity analysis; and
4) ground-water flow and contaminant transport modeling.
The 3-dimensional GIS must therefore interface with the remaining modules, for which many analytical tools have already been developed. Figure 8.4 attempts to summarize these concepts by showing the dominant information flows and cycles among these modules.

The process begins in the upper left corner of Figure 8.4, where the geologist investigator combines geological experience with limited field data to begin the subsurface characterization

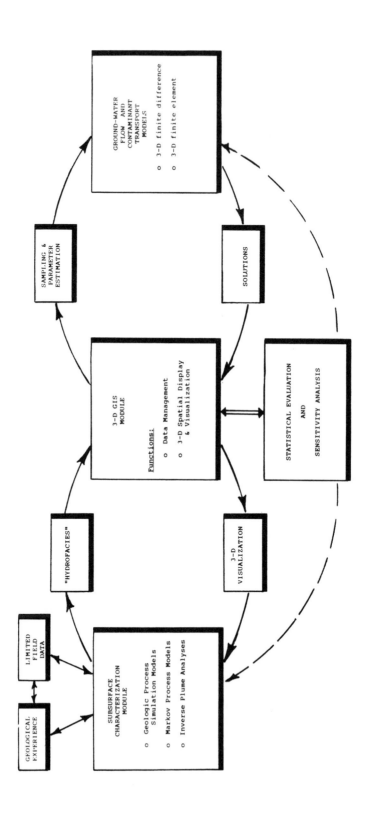

Figure 8.4 Information flow and the role of 3-dimensional GIS for hydrogeological applications.

process. The subsurface characterization module contains a variety of analytical techniques, including:

1) geological process simulation models which may combine both stochastic and deterministic elements;
2) one-, two-, or three-dimensional Markov process models; and
3) inverse plume analysis methods.

All these methods have been defined previously.

The information generated by thesubsurface characterization module is linked in a circular fashion to the 3-dimensional GIS module. This feed-back loop is an important consideration in defining appropriate linkages between the GIS and the analytical tools within the subsurface characterization module. It must be designed to provide both data management and spatial visualization support. A number of iterations are expected before the most probable subsurface conditions are defined. In some cases, a unique solution may not be achievable, and two, or more, alternative characterizations may be used.

Once a suitable subsurface characterization has been defined, the analysis now moves into a second cycle, shown on the right of Figure 8.4, where the 3-dimensional GIS interacts with appropriate ground-water or contaminant transport models. This involves the creation of finite difference or finite element meshes by sampling from the data base. The definition of an optimal mesh has recently been studied by the author and some colleagues (Stam et al. 1989).

Figure 8.4 shows a strong linkage between the 3-dimensional GIS module and a module labelled "Statistical Evaluation and Sensitivity Analysis". A full discussion of the analytical procedures located within such a module is beyond the scope of this presentation; only the basic functions of the module will be described.

From a practical and efficiency point-of-view, some method of assessing the "usefulness" and "reasonableness" of the subsurface characterizations must be produced. Without them, the investigator cannot make rational decisions concerning when the characterization is complete or adequate. Only when the subsurface conditions are clearly defined, can the investigation move into an analysis of the hydrogeological conditions.

The spatial visualization capabilities of the 3-dimensional GIS are obviously one way of making such an assessment. However, other more numerical approaches seem necessary. They would include standard statistical screening methods, but also more sophisticated "geostatistical" techniques. The term "geostatistics" has been used to describe several spatial interpolation methods using spatial autocovariance functions, the so-called "regionalized variables" (Olea 1975; Clark 1979; Lam 1983). These techniques, often referred to as "Kriging", have been widely applied within the geosciences. They assume the data are time-invariant. Another method, Kalman filtering, allows both time and spatial variation in the data (van Geer 1987). These methods have been used for the optimization of sampling networks, but appear to have special utility in analyzing seasonally varying contaminant data.

The use of such techniques, in conjunction with the analytical methods contained within the other modules, allow for a second level of information cycling and feed-back. This is shown by the dashed arrows in Figure 8.4. An important question that is often posed in ground-water contamination modeling studies concerns the sensitivity of the answers to variations or uncertainities in the input parameters. This "sensitivity analysis" requires the combined use of all the modules shown in Figure 8.4.

Conclusions

Three-dimensional GIS technologies have a central role to play in future quantitative assessments required by such hydrogeological applications as petroleum reservoir characterization and ground- water contamination modeling at hazardous waste sites. Three-dimensional GIS cannot solve these problems without closely interfacing with a variety of existing analytical techniques for subsurface characterization, ground-water modeling, and statistical assessment.

The functionality of future 3-dimensional GIS products must strike a balance between their dual roles of spatial visualization and data management. Recent speculation (Lang 1988) concerning the merging of GIS and CAD/CAM technologies emphasize the shortcomings of the current systems for many geoscience applications.

References

Allen, J. R. L., 1965, A review of the origin and characteristics of recent alluvial sediments. *Sedimentology*, **5**, 89-191.

Allen, J. R. L., 1974, Studies in fluviatile sedimentation: implications of pedogenic carbonate units, Lower Old Red Sandstone, Anglo-Welsh outcrops. *Journal of Geology*, **9**, 181-208.

Allen, J. R. L., 1978, Studies in fluviatile sedimentation: an exploratory quantitative model for the architecture of avulsion-controlled alluvial suites. *Sedimentary Geology*, **21**, 129-147.

Augedal, H. O., Omre, H., and Stanley, K. O., 1986, SISABOSA- A program for stochastic modeling and evaluation of reservoir geology, In Proceedings, *Reservoir Description and Simulation*. Institute for Energy Technology (IFE) Norway, Oslo, Norway, September 1986.

Belcher, W. R., 1988, Assessment of aquifer heterogeneities at the Hanford Nuclear Reservation, Washington, using inverse contaminant plume analysis. Colorado School of Mines Engineering Report ER-3594.

Bridge, J. S., 1979, A FORTRAN IV Program to Simulate Alluvial Stratigraphy. *Computers & Geosciences*, **5**, 335-348.

Bridge, J. S., and Leeder, M.R., 1979, A simulation model of alluvial stratigraphy. *Sedimentology*, **26**, 617-644.

Cant, D. J., and Walker, R. G., 1976, Development of a braided fluvial facies model for the Devonian Battery Point Sandstone, Quebec. *Canadian Journal of Earth Sciences*, **13**, 102-119.

Cant, D. J., and Walker, R. G., 1978, Fluvial processes and facies sequences in the sandy braided South Saskatchewan River, Canada. *Sedimentology*, **25**, pp.625-648.

Carr, D. C. et al., 1966, Stratigraphic sections, bedding sequences, and random processes. *Science*, **154**, 1162-1164.

Clark, I., 1979, *Practical geostatistics*. (London: Applied Science Publishers), 129p.

Davis, J. C., 1986, *Statistics and data analysis in geology*, 2nd ed, (New York: John Wiley and Sons), 646p.

Domenico, P. A., 1987, An analytical model for multidimensional transport of a decaying contaminant species. *Journal of Hydrology*, **91**, 49-58.

Domenico, P. A. and Robbins, G. A., 1985, A new method of contaminant plume analysis. *Ground Water*, **23**, 476-485.

Fogg, G. E., 1986, Groundwater flow and sand body interconnectedness in a thick, multiple-aquifer system. *Water Resources Research*, **22**, 679-694.

Fogg, G. E., and Kreitler, C. W., 1981, Ground-water hydrology around salt domes in the East Texas Basin: a practical approach to the contaminant transport problem. *Bulletin Association of Engineering Geologists*, **18**, 387-411.

Fogg, G.E., Seni, S. J., and Kreitler, C. W., 1983, Three-dimensional ground water modeling in depositional systems, Wilcox Group, Oakwood Salt Dome area, East Texas. *Texas Bureau of Economic Geology Report of Investigations* **133**.

Gaylord, D. R. and Poeter, E. P., 1988, *Annual report of geologic and geohydrologic site characterization studies during FY88, Hanford Reservation, Washington.* Battelle Pacific Northwest Laboratory, Richland, Washington.

Gelhar, L. W., and Axness, C. L., 1983, Three-dimensional stochastic analysis of macrodispersion in aquifers. *Water Resources Research*, **19**, 161-180.

Gupta, S. K., Cole, C. R., Kincaid, C. T., and Monti, A. M., 1987, *Coupled fluid, energy, and solute transport (CFEST) model: formulation and user's manual.* Battelle Memorial Institute, Office of Nuclear Waste Isolation, Report BMI/ONWI-660.

Haldorsen, H.H., Brand, P.J., and MacDonald, C.J., 1987, Review of the stochastic nature of reservoirs, presented at Seminar on the *Mathematics of Oil Production*, Robinson College, Cambridge University, July 1987.

Harbaugh, J. W. and Bonham-Carter,G., 1970, *Computer simulation in geology.* (New York: John Wiley and Sons), 575p.

Krumbein, W. C., 1967, FORTRAN IV Computer programs for Markov-chain experiments in geology. *Kansas Geological Survey Computer Contribution.* **13**.

Lam, N.S., 1983, Spatial Interpolation Methods: A Review. *American Cartographer*, **10**, 129-149.

Lang, L., 1988, Are CAD and GIS evolving toward the same answer? *Geobyte*, Nov. 1988, 12-15.

Lavenue, A. M. and Domenico, P. A., 1986, A preliminary assessment of the regional dispersivity of selected basalt flows at the Hanford site, Washington, USA. *Journal of Hydrology*, **85**, 151-167.

Lin, C. and Harbaugh, J. W., 1984, Graphic display of two- and three-dimensional Markov computer models in geology, *Computer Methods in the Geosciences* **2**. (New York: Van Nostrand Reinhold).

Leeder, M. R., 1978, A quantitative stratigraphic model for alluvium with special reference to channel deposit density and interconnectedness. *Canadian Society of Petroleum Geologists Memoir* **5**, 587-596.

Miall, A. D., 1973, Markov chain analysis applied to an ancient alluvial plain succession. *Sedimentology*, **20**, 347-364.

Miall, A. D., 1977, A review of the braided river depositional environment. *Earth Science Reviews*, **13**, 1-62.

Miall, A.D.(editor), 1978, Fluvial sedimentology. *Canadian Society of Petroleum Geologists Memoir* **5**.

Olea, R. A., 1975, Optimum mapping techniques using regionalized variable theory. *Kansas Geological Survey Series on Spatial Analysis* **2**.

Powers, P. W. and Easterling, R. G., 1982, Improved methodology for using embedded Markov chains to describe cyclical sediments. *Journal of Sedimentary Petrology*, **52**, 913-923.

Smith, D. R. and Paradis, A. R., 1989, Three-dimensional GIS for the earth sciences. (this volume).

Stam, J. M. T., Zijl, W., and Turner, A. K., 1989, Determination of hydraulic parameters from the reconstruction of alluvial stratigraphy. Proceedings, *4th International Conference on Computational Methods and Experimental measurements*, Capri, Italy, (in press).

Walker, R. G.(editor), 1984, Facies Models, 2nd Edition,. *Geoscience Canada Reprint Series* **1**, Geological Association of Canada, Toronto, Canada.

van Geer, F. C., 1987, *Applications of kalman filtering in the analysis and design of groundwater monitoring networks.* TNO Institute of Applied Geoscience, Delft, The Netherlands.

Chapter 9

Spatial data structures for modeling subsurface features

Carl Youngmann

Introduction

An important step in the integration of geophysical and geological analyses is the creation of a comprehensive model of the earth's crust. This model must accommodate a variety of data inputs including both raw data and interpretations from wells, seismic surveys and other measures. The usefulness of the model comes from the ease with which it can be developed, modified and analyzed. A common integration method for modeling is to combine surface maps with fence diagrams. Alternative solutions have been sought in geometric solids modeling and topological volume models. The goal is to identify an editable and extendible model that encompasses the structure of the original information and permits the generation and re-integration of auxiliary analyses.

Earth Data

A comprehensive model of the earth's crust must accommodate both raw information and interpretations. Raw data may come from a variety of sources including well cores, well logs, and seismic surveys, as well as other measures such as gravity and magnetic studies. Interpretations may be in the form of sampled locations, well log blockings, cross-sections, grids of horizons, layers or faults. Accommodation of discrete and continuous attributes for curves and layers is needed. Surfaces and layers may represent the bounds of geologic or seismic facies resulting from deposition or structural tectonics. Well data may include a variety of geophysical, petrochemical, and geological curves. Positional information may be referenced to global latitude or longitude or rectilinear map grids with depths represented by time or actual depth.

The purpose of a geologic model is the description of the different rock formations, their characteristics and the interfaces between them. From this understanding of the gross physical structure and the physical and chemical composition of the component rocks, estimates of recoverable petroleum and mineral resources can be made. As with the creation of 2-dimensional, planimetric map models of spatial phenomena, the utility of a model comes from the ease with which it can be developed, modified and analyzed. The tools developed for these tasks must reflect the available data sources, the methods of interpretation to be applied, and the data requirements of analytical, reconstruction, and simulation methods. At any given time, the model must accommodate the extraction of interface layers, structural faults, cross-sections, time or depth layers, simulation blocks, or pseudo well cores. The model should form the central repository for an evolving interpretation of the earth's crust.

The traditional data sources for subsurface models are seismic cross-sections, well logs, well cores and surface structures. Raw seismic data collected in the field is subjected to massive processing regimes in order to derive vertical seismograms of signal amplitude. Interpretations of these seismograms include definitions of unconformities that may be interpreted as erosional truncations, depositional horizons, faults or formation contacts. In addition to picking the bounds of transition, zones can be interpreted into seismic facies of similar signal characteristics. These analyses may be done in a time domain or the seismograms may be migrated to depth through the application of velocity fields.

Well data comes from rock cores collected during drilling or from the logs of surveys conducted after completion with a variety of down-hole tools. Well information provides the only "ground truth" for adjusting seismic data; it is invaluable in determining the mineralogical attributes of a geologic model. Cores and logs are analyzed simultaneously in order to derive interpretations of the variation in rock types and their physical and chemical characteristics. Data for groups of wells are correlated between wells to develop interpretations of horizons, faults, and lithofacies. A variety of characteristics may be assigned to the bounding horizons and faults, as well as the resulting facies.

The inter-relationship of seismic and well data requires the correct relative positioning of this information. Directional surveys are used to determine the true subsurface location of well paths and the proper positions for log and curve data. Correspondence to seismic data may require depth migration of the seismic time information or unmigration of depth-measured well data to time. Problems of mis-alignment and measurement variation make the task of consolidating source information difficult.

Map Models

Creation of an integrated and consolidated interpretation drawing from the different data sources available is a major problem. Often the original data is never combined and only interpretations are associated. A commonly used method of integration is to combine surface maps with fence diagrams (Hamilton et al. 1987). This may be seen as a stack of maps (Figure 9.1) or a set of vertical slices (Figure 9.2). Usually, there are only implicit

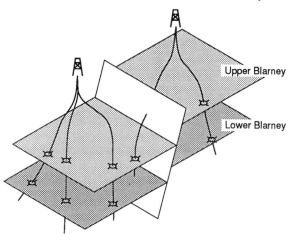

Figure 9.1 Normal fault showing intersecting surfaces

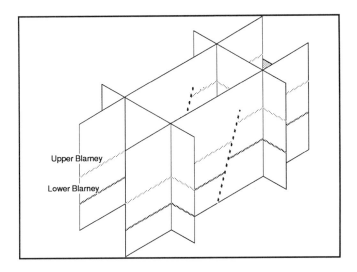

Figure 9.2 Fence diagram cross section model of a normal fault

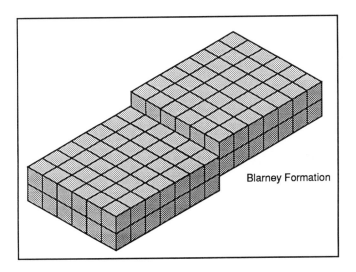

Figure 9.3 Reservoir modeling blocks extracted from intersecting surfaces interpreted as a layer

connections between the surfaces and horizons. Re-interpretation of an individual surface or cross-section may not result in adjustment of the collateral representation. The stack of maps and fence diagrams may be converted into other forms such as blocks (Figure 9.3), i.e. 3-dimensional grid cells for simulation studies in reservoir analysis (Jones & Bush 1987). In a conversion, the interface information maybe lost within the blocks, although the extraction of attributes within layers is facilitated.

Carl Youngmann

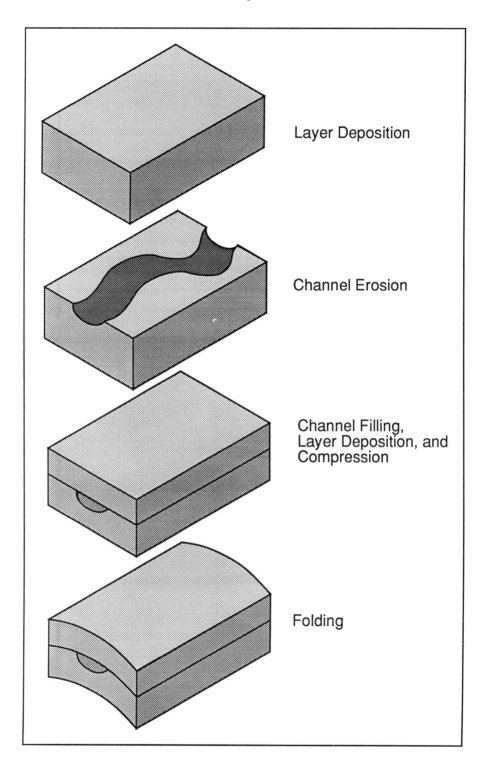

Layer Deposition

Channel Erosion

Channel Filling,
Layer Deposition, and
Compression

Folding

Figure 9.4 Solid model development showing depositional plane, channel erosion, deposition, compression, folding and fracturing.

Solid Models

An alternative solution for an integrated model has been sought in applications of geometric solids modeling methods used in mechanical design. The earth's crust can be viewed from a geologic process point-of-view in which layers are aggregated in a regular manner and erosional and tectonic forces are applied to modify these regular layers (Figure 9.4). A realistic geologic model, for example, can be developed from solids by starting with a simple set of layers and then applying forces to rotate, skew, distort and even shear the layers. Erosional truncations and incisions can be implied by subtractions and additions of other solids. However, there is dissatisfaction with this approach because of the tremendous number of assumptions and the greater understanding necessary to identify the generating processes. It seems that the production of a solid model might provide additional insight after the development of an integrated geologic model, but not as the integrated model.

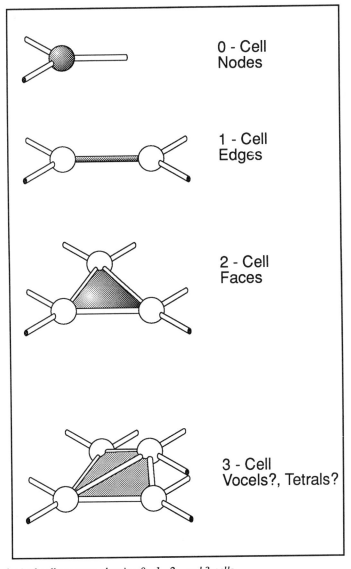

Figure 9.5 Topological cell concepts showing 0-, 1-, 2-, and 3-cells

Topological Models

A fully topological volume model that can be developed from surfaces and cross-sections offers some promise for integration. The geologic structure can be viewed as a closed, three-dimensional graph enveloping spaces of homogeneous geologic character. The basic concept of a map as the topological combination of cells can be extended from 0-cell points, 1-cell arcs, and 2-cell regions to include 3-cell interstices (Figure 9.5). The methods for such graph creation and inversion are well-known for single valued maps (Corbett 1979). Their application to multi-valued surfaces and volumes is less well-developed (Frank and Kuhn 1986; Carlson 1987).

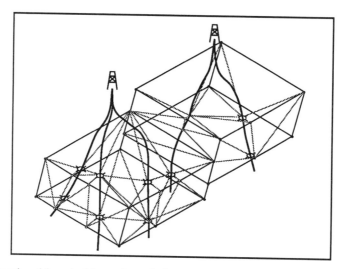

Figure 9.6 Structural model emphasizing nodes and edges.

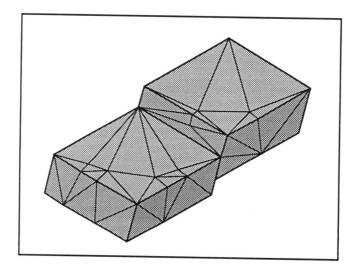

Figure 9.7 Tesselated tetrahedral model.

The polyhedral topology can be structured to encompass the positional and attribute information about the interfaces and spaces (Figure 9.6). However, a major question is whether the information should be structured into a 3-dimensional tetrahedral tesselation or should the data be in 3-dimensional patches in a boundary model. In a tetrahedral tesselation, the volume is divided into 4-sided irregular solids of homogeneous characteristics (Figure 9.7). With 3-dimensional patches, 3-cells are created from non-linear 2-cells the shape of which is defined either parametrically by surface equations or deterministically with data points (Figure 9.8).

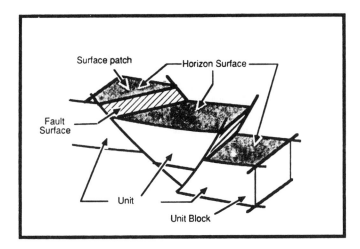

Figure 9.8 Topological model with deterministically specified surface patches.

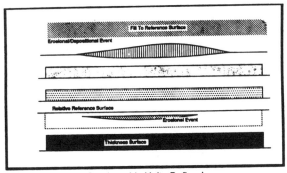

Stratigraphic Units Defined

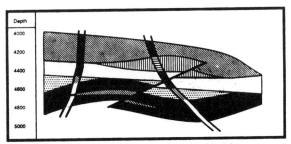

Final Structural and Stratigraphic Model

Figure 9.9 Recumbent and reverse folding as sources of multi-value problems in surface specification.

Difficulties in the use of the topological model for the integration of geological, geophysical and petrophysical data come in the need to maintain the relationships between interpreted surfaces and faults. An easily edited model would provide the ability to pick bounding horizons with intervening layers that are discontinuous due to both faulting and folding (Figure 9.9). This situation leads to multi-valued surfaces that can not be constructed from standard gridding procedures; it must be modeled by Coombs patches or nerve surfaces. Furthermore, there exists the necessity to be able to follow disconnected surfaces over intervening fault planes with the ability to create and remove fault planes dynamically.

Conclusion

The topological model appears to offer the greatest potential for encompassing the diversity of the data, but it requires greater complexities in the formation and maintenance of the model. However, the solid model most accurately reflects the view of the reservoir analyst, but they are difficult to express and pose many problems for editing. On the other hand, the overlain surfaces and intersecting cross-sections of a map model represent the forms of the data that most analysts are familiar with, but they present numerous problems in integration and editing.

References

Carlson, E., 1987, Three-dimensional conceptual modeling of subsurface structures, *Technical Papers*, **4**, ASPRS-ACSM Annual Convention, Baltimore, Maryland.

Corbett, J.P., 1979, *Topological Principles of Cartography*, Technical Paper **48**, Bureau of the Census.

Frank, A. U. and Kuhn, W., 1986, Cell graphs: A provable correct method for the storage of geometry, *Proceedings, Second International Symposium on Spatial Data Handling*, Seattle, Washington.

Hamilton, D. E., Downing, J. A., and Jones, T. A., 1987, Graphic interface for a geologic mapping system, *Proceedings, NCGA Mapping and Geographic Information Systems '87*, San Diego, California.

Jones, T. A. and Bush, M. M., 1987, Graphical display of three-dimensional geologic Models, *Proceedings, NCGA Mapping and Geographic Information Systems '87*, San Diego, California.

Chapter 10

Creating a 3-dimensional transect of the earth's crust from craton to ocean basin across the N. Appalachian Orogen.

John D. Unger, Lee M. Liberty, Jeffrey D. Phillips, and Bruce E.Wright

Introduction

The Quebec-Maine-Gulf of Maine Global Geoscience Transect extends from the stable Precambrian platform of North America across the Northern Appalachian orogen to the Atlantic Ocean basin south of Georges Bank. This project is the first of its kind to attempt to characterize a volume of the earth's crust the dimensions of a transect (880 km long, 100 km wide, and up to 45 km deep) using 3-dimensional, digital modeling techniques.

We approached the problem by dividing parts of the transect into discrete blocks approximately 50 km by 70 km in areal extent and extending down to the base of the crust, or MOHO. These blocks are being used to construct 3-dimensional geologic models with Dynamics Graphics' Interactive Surface Modeling (ISM) software. The relatively straightforward approach of using surfaces to represent the third dimension, rather than to create a true 3-dimensional volume with cubes or voxels that contain x, y, z, and attribute information, makes the computational aspects of the model simpler. However, this approach may not allow easy transportability of the data to state-of-the-art hardware and software that can display and manipulate 3-dimensional models.

Five principal geological/geophysical data sets were combined with a digital elevation model and used to construct our 3-dimensional models:

1) migrated seismic reflection profiles;
2) seismic refraction data;
3) gravity models;
4) magnetic models; and
5) a bedrock geological map.

These five sources of information were blended together as described in this paper to construct the most accurate and reliable 3-dimensional models of the crust. These models yield unprecedented detail about the upper 10 km of the crust and show broad outlines of structures within the deep crust. The surface of the MOHO can be contoured with approximately ±2 km uncertainty.

The United States Geological Survey (USGS) with the cooperation of the Geological Survey of Canada (GSC) is constructing a digital geological/geophysical transect (NA-1) which is part of the Global Geoscience Transect Project of the International Commission on the Lithosphere. The transect begins near Quebec City on the edge of the North American

Precambrian craton, enters the Appalachian orogen in southern Quebec, continues across this mountain belt in Maine and extends offshore across the Gulf of Maine, and ends just southeast of George's Bank at the edge of the Atlantic Ocean basin (Figure 10.1).

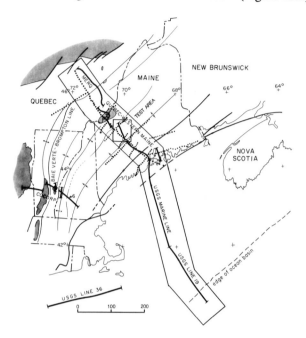

Figure 10.1 Map of the Northeastern United States and adjacent Canada showing some of the major geological features of the region. The patterned areas are outcrops of Grenvillian basement rocks. Also shown are the transect boundaries (the irregular polygonal area), location of test area 1 in central Maine (the rectangle in central Maine that lies inside the transect), the seismic reflection lines (the heavy lines inside and outside of the transect area), and the seismic refraction deployments and fans (the dotted lines).

All of the major data sets that comprise this transect have been digitized, and it is our goal to produce a 3-dimensional digital model of the crust along the transect corridor 880 kilometers long, 100 kilometers wide, and up to 45 kilometers deep. In this model we are striving to depict the major geological and geophysical boundaries present in the earth's crust to the greatest detail possible given the limitations of our knowledge and the capabilities of the computer tools we are using.

Data

The composite 3-dimensional modeling that will be discussed is derived from interpretations and models generated by five data sets: the bedrock geology, the seismic reflection and seismic refraction data, and by the gridded Bouguer gravity and aeromagnetic data. To begin, the individual data sets will be described, and then we will discuss how they are interpreted and integrated to construct our model.

The primary data set for any project such as ours must be the surface bedrock geology. This information provides "ground truth" for the other data that are used to build the model and is the 2-dimensional foundation upon which the rest of the model must be constructed. The geological data for our transect are derived from three major sources: for Maine from

Osberg, et al., (1985); for southern Quebec St. Julien and Slivitsky (1985) and Globensky (1985); and for the Gulf of Maine (Klitgord, et al., 1988). The maps for Quebec and Maine are traditional, hand- produced, printed geologic maps, which had to be digitized before they could be used in our study. The geologic data for Maine were digitized by scanning the Maine state geologic map with SCITEX hardware and software at the USGS. The raster data generated by the SCITEX process were converted into a vector format, Digital Line Graph (DLG), and transferred in that format to ARC/INFO, a GIS software package marketed by Environmental Systems Research Institute. The Quebec geologic data were first scribed by personnel at the GSC and then machine digitized and converted into vector format with the ARC/INFO software. The Gulf of Maine geology was digitized by hand by USGS personnel and converted to ARC/INFO format. All of the digitized geology and other geologic features within the transect area were given DLG attribute codes in ARC/INFO according to a coding scheme devised at the USGS by B. E. Wright and D. B. Stewart (in preparation).

The geology along our transect is extremely complex and consists of folded and faulted blocks of rocks that are separated by high angle thrust, normal, and strike-slip faults and by low angle detachment faults with large offsets. These rocks are also cut by large and small bodies of intrusive igneous rocks. Our knowledge of the geology is much more complete on land than in the offshore portion of the transect, though the offshore geology is thought to be a continuation of the rocks exposed onshore.

In trying to determine the best methods and procedures to work with the geology and with the other data sets within the transect, we chose test areas within the transect boundaries that contained all of the representative data sets and where we could experiment with the techniques we plan to use for the transect as a whole. The location of test area 1 is shown in Figure 10.1, and the remainder of this paper will focus on the results of our efforts to date with modeling the crust in this 50 x 70 km area, which is shown in that figure and in more detail in Figure 10.2.

The important fact to keep in mind is that the level of detail digitized for the onshore bedrock geology is greater that we can model in the subsurface and that we must consider a simpler, more generalized representation of the geology when creating the 3-dimensional model. Therefore, the next step in our analysis was to transfer the complete geologic information in DLG format from ARC/INFO to the 3-dimensional modeling software we have been using, Interactive Surface Modeling (ISM) marketed by Dynamics Graphics Corporation, and to simplify it in that environment. Although all the major geological features were retained, many of the smaller units were grouped into larger entities by generalizing the bedrock geology and editing the data after they had been transferred to ISM and incorporated in that package as a surface annotation file (Figure 10.2).

The major geologic units are shown by the simplified geologic map in Figure 10.2. The Siluro-Devonian Kearsage-Central Maine synclinorium covers most of the test area and consists of pelites, sandstones, and thin limestones. The rocks of the synclinorium were tightly folded parallel to the northeast- southwest axis of the synclinorium and then metamorphosed from greenstone to sillimanite grade. These units are bounded on the northwest by older, folded and faulted Cambrian – Ordovician metasedimentary strata that are underlain by an ultramafic complex and a different, non-Grenvillian Precambrian basement (the Chain Lakes Massif) that was thrust onto the North American continental plate during the Taconic orogeny. All of the rocks have been intruded by early Devonian plutons such as the granitic Lexington batholith shown in the center of Figure 10.2 and the diorites and gabbros of the Flagstaff Lake and Pierce Pond plutons, which lie west and north of the Lexington batholith, respectively.

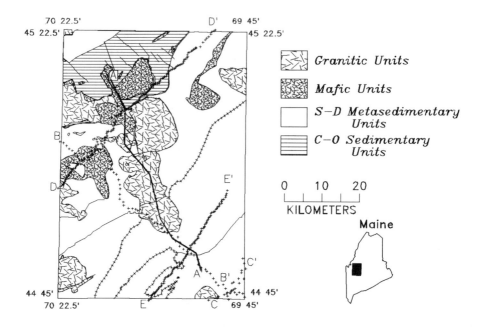

Figure 10.2 A simplified geologic map of test area 1 in the Quebec– Maine– Gulf of Maine transect. The reconstruction line for the reflection profile that passes through this area is shown by A– A'. The alignments B– B' and C– C' show the portions of the refraction deployments and profiles that lie in the test area. D– D' and E– E' are the projected subsurface reflection points of the refraction fan deployments.

The geologic data are used to give the viewer a visual tie with those elements of the model that intersect the earth's surface. Therefore, in our model the simplified bedrock geology is displayed by being draped over the gridded surface elevation data. Of course, information such as strikes and dips of the major geologic structures are used to project the bedrock geology into the third dimension, especially in the shallowest parts of the model and where the geophysical data are ambiguous or absent.

Within the transect area as a whole we have good coverage with seismic reflection and refraction data (Figure 10.1). Test area 1, shown in Figure 10.2, contains a major deep reflection profile, A-A', sections of two refraction profiles, B-B' and C- C', and parts of two refraction fans, D-D' and E-E'. The acquisition and processing of the reflection and refraction data are described by Stewart et al., (1986); Unger et al., (1987); Luetgert et al., (1987); and Spencer et al., (1988). The accurate location of each of these data sets is vital for constructing a sensible and consistent model. The principal type of information that we get from the seismic data is a profile or pseudo-cross section through the crust. For the reflection lines, we use common mid-point (CMP) stacked data that have been migrated to display the major reflectors seen along the profiles in their correct relative positions. Migration is a computational procedure to convert reflected seismic energy from its original measurements in two-way travel time to distance and to correct the positions of dipping reflectors (Unger, 1988). We then strive to simplify the patterns of migrated reflectors by interpreting which of them represent major crustal boundaries using the character of the reflected energy, the strength of the reflectors, and the velocity deduced during data processing. The interpreted major boundaries, such as the MOHO, are then used to construct or constrain the 3-dimensional model. However, we retain the ability to project the original migrated data on the model after it is constructed to make sure that we haven't taken too many liberties with our interpretations. An example of this procedure can be seen on Figure 10.7, where a cross-section through the model was chosen to pass through the same surface locations as the seismic reflection profile, and the reflectors from the profile can be seen overlying the other components of the model.

For the refraction data, three separate studies were carried out:

1) Ray tracing modeling was done in the upper 10 kilometers of the crust along all of the refraction deployments (Luetgert, et al., 1987). These studies define the shallow velocity structure of the materials beneath the line (Figure 10.3a);

2) Reflected arrivals from deeper surfaces seen from wide-angle reflections in the normal refraction deployments were examined and processed using normal moveout techniques similar to those used for high-angle reflection data analysis;

3) Wide-angle reflections from "fan" deployments (Klemperer and Luetgert, 1987) where the signal source was located perpendicular to the alignment of recording instruments, were also processed using normal moveout techniques to define mid- and deep- crustal reflector surfaces such as the MOHO (Figure 10.3b).

Measurements of the gravity field that had been collected throughout the transect area were compiled (Stewart, et al., 1986), and combined with new gravity measurements (Phillips, et al., 1988). We took advantage of the accurate elevations and locations computed for the seismic reflection stations to make new, closely spaced gravity measurements along all of the reflection profiles. The gravity data were then processed and the results interpolated and extrapolated to form a 2-dimensional grid, with 0.5 km spacing, of Bouguer gravity anomaly values over the transect and test areas (Figure 10.4a).

Aeromagnetic data from flight lines with spacings ranging from 0.25-mile to 3-miles were processed and used to construct gridded anomalies at the same 0.5 km spacing as the gravity anomaly grid (Figure 10.4b).

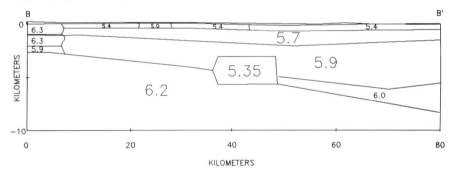

Figure 10.3a A velocity model of the upper 10 km of the crust which has been derived from seismic refraction ray tracing along deployment B - B' (Figure 10.2). The numbers inside the elements of this 2-dimensional model are compressional wave velocities in km/sec (after Luetgert, et al., 1989).

One way in which we have used these gravity and aeromagnetic data is to develop 2.5-dimensional cross sections, from both measured and gridded data, along profiles at a regional scale that cross large parts of the transect as well as across smaller areas such as test area 1 (Figure 10.5). (The process of computing the 2.5-dimensional potential field cross sections takes into account rocks lying outside the plane of the section and assumes no changes in the geologic structure in this direction. See Shuey and Pasquale (1973) for further discussion on this method). We can construct these cross sections through the parts of the area where we perceive the most interesting geologic problem exist, or we can make cross sections that coincide with the reflection or refraction profiles in order to compare and confirm our models.

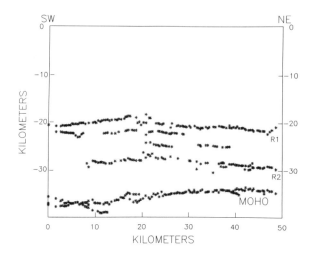

Figure 10.3b A profile of the strongest wide angle reflections picks from refraction fan 2 (C - C', Figure 10.2). The crosses that are reflections from the MOHO plot across the bottom of the figure; those labeled R1 and R2 represent typical reflections from horizons in the deep crust (after Luetgert et al., 1989).

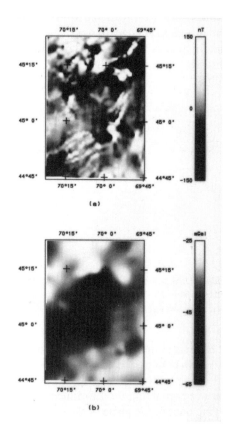

Figure 10.4 Aeromagnetic (a) and Bouguer gravity (b) fields for the test area gridded at 0.5 km and displayed as grey scale images.

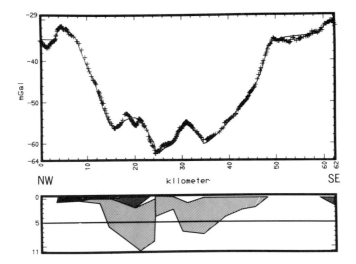

Figure 10.5   The results of gravity modeling along the profile A– A' in Figure 10.2.   The crosses in the upper part of the figure show the Bouguer gravity as measured along A– A'.   The solid line in the upper portion of the figure shows the calculated Bouguer gravity that results from 2.5-dimensional modeling of the body shown in the lower part of the figure.   The lighter pattern in the model represents granitic rocks with an average density 0.17 gm/cc less than the surrounding strata, and the darker pattern mafic intrusives with an average density 0.12 gm/cc greater than the surrounding rocks.

A second way that the gridded data are interpreted is by applying a 2-dimensional filter to the data to determine how the spatial frequency components of the potential fields are related to the observed geologic features (Figure 10.6). This process gives us insight into the thicknesses and shapes of the intrusive bodies, shows us the subsurface orientation of complexly folded magnetic metasedimentary rocks, and provides a generalized interpretation or image of how geologic structures vary with depth. For example, the strongest anomalies seen in the short wavelength magnetic map, but absent in the long and intermediate wavelengths, are most likely due to shallow geologic features.

Modeling procedure and a test model

Our final task was to put the information from the five different data sets together to form a crustal model of test area 1. Our emphasis in this process was to create the surfaces of the model in a consistent manner by interpolating and extrapolating our interpretations where required. These operations allow us to compare and contrast features or surfaces that are defined by two or more of the data sets.

Where we have overlapping data sets we could directly compare the elements of our model with the original data used to define surfaces in the model. For example, as mentioned above, along the trace of the seismic reflection profile, detailed closely spaced gravity measurements have made it possible for us to construct a 2.5-dimensional gravity model. We can now compare that model directly with the reflectors seen in the migrated stacked seismic reflection profile (Figure 10.7). These procedures enabled us to model many of the geologic structures in the upper 10 kilometers of the test area.

One of the best-defined interfaces seen in our data is the MOHO. This boundary was well imaged by both the high angle seismic reflection survey and by the wide angle reflection data

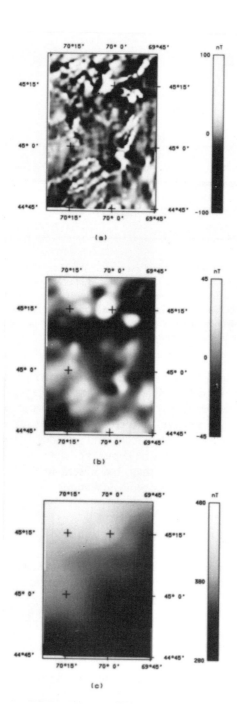

Figure 10.6 Bandpass filtered aeromagnetic data for the test area. Highpass filtering (a) emphasizes narrow features produced by shallow sources. Intermediate bandpass filtering (b) emphasizes the broader, more general features of the mapped bedrock geology. Lowpass filtering (c) emphasizes the regional magnetic field, which in this case mimics the surface of the MOHO.

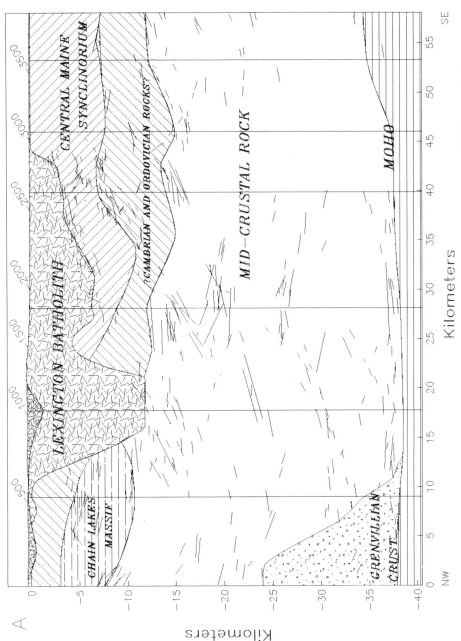

Figure 10.7 A cross-section taken through our 3-dimensional model shown in Figure 10.8. This section is oriented along the reflection profile A - A' (Figure 10.2). Superimposed on the cross-section are the migrated reflectors from the seismic reflection profile. The continuous lines are surfaces from the model and represent major geological boundaries.

from the refraction experiment. Where the data sets coincide, these independent measurements produce a consistent crustal thickness. The general shape and orientation of the MOHO surface is also evident in the long wavelength portion of the filtered aeromagnetic data (Figure 10.6c).

To construct a 3-dimensional model using the ISM software, one must first define a set of interfaces or surfaces, which in our model correspond to the contacts between geologic units. To define a surface, ISM is given a set of points in X, Y, Z space that lies on some boundary.

One then uses ISM to take these data and, using a bicubic spline algorithm, construct a continuous grid or surface of Z values at a uniform X, Y spacing for the boundary. Obviously, the quality of such a gridded surface depends on the density of the data points used to define it. Some of the surfaces used in our model, such as the MOHO and the bottom of the Lexington batholith, are relatively well defined by more than one data set. Others, such as the bottom of the Kearsarge-Central Maine synclinorium are less well known and have been extrapolated from sparser data. ISM supports grid sizes up to an X, Y dimension of 512 x 512. For our data sets, we have found that a square grid with a 500 meter spacing is sufficiently detailed to support our modeling effort and large enough to display a reasonable part of the transect with a single model. One ISM model can contain up to 12 surfaces, which can be displayed together in a variety of ways. For example, one can construct an arbitrary cross-section through the model (Figure 10.7) or make a fence diagram and display many cross-sections. Alternatively, the model can be viewed from an arbitrary perspective as a block diagram (Figure 10.8).

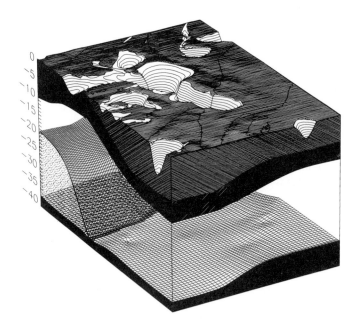

Figure 10.8 A perspective view of the 3-dimensional ISM model constructed for test area 1. This model shows the major interfaces currently defined for this volume of the crust. The upper surfaces of the intrusive igneous rocks have been "stripped" off so that the intrusives appear as holes contoured at 1 km depth intervals. The lowest surface is the MOHO; the surface that dips steeply from the back (north side) of the model and then terminates against the MOHO is the buried edge of Grenvillian crust.

New directions

There appear to be two key steps to accomplishing our goal of creating a true 3-dimensional model of the earth's crust: first we have to construct our model using the geological and geophysical data described above. Second, we must display this model in the most revealing and flexible way. At the present stage of development of 3-dimensional software, no single package is available that can do both of these tasks. Therefore, our approach has been to use ISM to construct the model using our digital data sets and our knowledge of the region. We can do a certain amount of limited display of the data using ISM, in fact, many of the final figures in this paper were prepared using ISM. However, the display capabilities of ISM are limited compared to other, state-of-the-art software and hardware systems.

One problem with ISM is that only 3-dimensional surfaces within an arbitrary rectangular volume can be defined. Information about the rocks between two surfaces is conveyed only by inference. It is this fact that distinguishes ISM from a true 3-dimensional volume modeling program. Our present task is to take the multiple surface model we have defined with ISM and convert it into a voxel model where each of the individual cubes or voxels that makes up the discrete elements of a volume can be assigned various attributes.

One hardware/software system that possesses powerful capabilities to display and manipulate true 3-dimensional data in this form is the PIXAR image display workstation. The USGS has recently acquired a PIXAR and our current research efforts center around using this system to display models such as that shown in Figure 10.8 of test area 1. We ultimately hope to be able to use the PIXAR system to view the digital model of the entire transect. Currently, the PIXAR system does not have a good method to construct a 3-dimensional model directly from data sets such as ours, so we will be forced to rely on vector or surface- oriented software like ISM to put our data into a model and then convert these ISM data to a voxel-based data set where they can be displayed and manipulated with the PIXAR or with another, comparable systems.

References

Globensky, Y., 1985, *Basses-Terres du Saint-Laurent, Quebec*, Map **1999** and Report **85-02**, Ministere de L'Energie et des Ressources, Quebec, Quebec.

Kiltgord, K. D., Hutchinson, D. R., Bothner, W. A., Trehu, A. M., McNab, R., Wade, J. A., and Keen, C. E., 1986, *Gulf of Maine: Tectonics and magnetic features*, 1986 draft.

Klemperer, S. L. and Luetgert, J. H., 1987, A comparison of reflection and refraction processing and interpretation methods applied to seismic refraction data from coastal Maine, *Bulletin of the Seismological Society of America*, **77**(2) 614-630.

Luetgert, J. H., Mann, C. E. and Klemperer, S. L., 1987, Analysis of deep crustal reflections from seismic refraction: data in the northern Appalachians, *Geophysical Journal of the Royal Astronomical Society*, **89**(1).183-188.

Luetgert, J. H., W. E. Doll, and Murphy,J., 1989, Seismic refraction profiles in northeastern Maine, to be submitted to *Bulletin of the Seismological Society of America*.

Osberg, P. H., Hussey, A. M., and Boone, G. M., (eds.), 1985, *Bedrock geologic map of Maine*, Maine Department of Conservation, Maine Geological Survey, 1 sheet.

Phillips, J. D., Thomas, M. D., and Jahrling, C. E., 1988, Principal facts for gravity stations along lines 1, 2, 3A, and 3B of the Quebec-Western Maine seismic reflection profile and among MERQ seismic reflection line 2001, Southeastern Quebec and west-central Maine, U.S.G.S. *Open- file Report* **88-425**, 46 pp.

St. Julien, P. and Slivitsky, A., 1985, L'Estrie-Beauce, Quebec, Map **2030** and Report **85-04**, Ministere de L'Energie et des Resources, Quebec, Canada.

Spencer, C. P., A. Green, J. H. Luetgert, P. Morel, D. B. Stewart, and J. D. Unger, 1988, The interpretation of refraction and reflection data from southeast Quebec, submitted to *Tectonophysics*

Stewart, D. B., Unger, J. D., Phillips, J. D., Goldsmith, R., Spencer, C. P., Green, A. G., Loiselle, M. C., and St. Julien, P. 1986, The Quebec-Western Maine Seismic Reflection profile: Setting and First Year Results. *American Geophysical Union, Geodynamics Series*, **14**, "Reflection Seismology: The Continental Crust", pp 189-199.

Shuey, R. T., and Pasquale, A. S., 1973, End corrections in magnetic profile interpretation, *Geophysics*, **38**(3), 507-512.

Unger, J. D., Stewart, D. B., and Phillips, J. D., 1987, Interpretation of migrated seismic reflection profiles across the northern Appalachians in Maine, *Geophysical Journal of The Royal Astronomical Society*, **89**(1), 171-176.

Unger, J. D., 1988, A PC program for migration of seismic reflection profiles, *GeoByte*, **3**, 42-54.

Wright, B. E., and Stewart, D. B., 1989, Digitization of a geologic map for the Quebec-Maine-Gulf of Maine Global Geosciences Transect, *American Cartographer*, in preparation.

Chapter 11

Three-dimensional GIS for the earth sciences

Dennis R. Smith & Arthur R. Paradis

Introduction

Different application groups have been using Geographic Information Systems (GIS) for various reasons. In talking to these people you soon realize that GIS means different things to different people. Over the years there have been attempts to define what a GIS is and how it is used; recent GIS reviews include Cowen (1988) and Clarke (1986). For the purposes of this paper we will use the following definition of a GIS. A GIS is a software system that contains functions to perform input, storage, editing, manipulation, analysis and display of geographically located data.

Up to now the main uses of a GIS have dealt with data on the earth's surface. If the data was above or below the surface it was conveniently projected to the surface. This allowed the system to deal with everything in a 2-dimensional format. Early GIS's often used a data structure of regular grid cells but current systems seem to favor polygons. All of these deal with many flat files that are oriented over the same location of the earth. Sometimes a system could draw a perspective view of the surface and even present data on the surface, below the surface and above the surface. These presentation techniques still deal with flat files but add the capability to present the data in what we will call a 2.5-dimensional format.

Why three-dimensional geoprocessing?

Many earth scientists who have tried their hand at geoprocessing have come up short. The use of flat, 2-dimensional files does not fit their needs. These scientists are usually dealing with geology, geophysics, meteorology, hydrology, mining, ground water, hazardous contaminations, and the like. These phenomena are 3-dimensional in nature and when you try to fit them into 2-dimensional systems you can not accurately model, analyze or display the information.

To help explain things throughout this paper we will use an example from a situation that we all hear about these days; the problems with hazardous chemicals in the ground. At this particular site they discovered, in the ground, PCB concentrations that were above safe levels. This indicated to the site owner that an expensive undertaking was necessary to first determine the extent of the problem and then to correct the situation.

It is impossible to model, analyze or display this situation, with any satisfaction, when you are using a 2-dimensional tool. You might be partially successful in using stacked 2-

149

dimensional data layers but you are basically forced to ignore the fact that the phenomena are actually 3-dimensional. Applying a 2-dimensional tool to 3-dimensional situations limits the scientist's work in many ways. It is not possible to accurately model the vertical relationships between the stacked 2-dimensional layers or to perform true 3-dimensional analytic operations between different models. Neither is it possible to accurately visualize the 3-dimensional situation and make decisions about the data.

Three-dimensional data

Let's go to our example site and take a look at the source data that is available. Contamination was discovered in the ground and new wells were drilled to gather additional data. Samples were taken at various locations down these wells and sent to the lab for analysis, where high levels of PCB were discovered. The site was therefore highly contaminated and had to be cleaned up. The PCB values were reported in a tabular fashion with the geographic location of each sample. With a 3-dimensional data set you need to know X, Y, Z & V where X, Y , Z give the location of the property, and V is the value of the property at that location. The property in this situation is the concentration level of PCBs. A portion of the data file is shown in Table 11.1.

Well-ID	X-coordinate	Y-coordinate	Elevation	PCB-level
2002	-1165	763	-80	0.33
2002	-1165	763	-140	0.16
2002	-1165	763	-200	0.66
2003	-1140	743	-20	0.05
2003	-1140	743	-80	0.06
2003	-1140	743	-140	0.09
2003	-1140	743	-200	0.13
2004	-1165	718	-20	0.13
2004	-1165	718	-80	0.45
2004	-1165	718	-140	0.10
2005	-1200	743	-20	0.13
2005	-1200	743	-80	0.72
2005	-1200	743	-140	0.09
2005	-1200	743	-200	0.33
2006	-1175	600	-20	0.19
2006	-1175	600	-80	0.22
2006	-1175	600	-140	0.14

Table 11.1 Portion of the Source File

With a 2-dimensional system the X & Y location of each well can be displayed and selected horizontal planes can be utilized in an attempt to model, analyze and display slices through the earth. With a 3-dimensional system the data is input, edited, modeled, analyzed and displayed in its true 3-dimensional form. Plate IX shows a display of source data with a 3-dimensional system.

Three dimensional modeling

In 2-dimensional systems the user often models the scattered or randomly located source data into a uniform or regular data structure. These have typically been uniform grids or triangles. The reason for this modeling is that the resources required to analyze and display data that is located in a scattered format is significantly higher than dealing with data in a regular format. Each time the scattered data needs to be contoured, or each time volumes need to be calculated, a modeling step would have to take place. The intention is to perform the modeling step once. As long as the mathematical model fits the physical model, the savings in resources are valuable.

The situation with 3-dimensional phenomena is similar. It is more efficient to model the scattered data once onto a uniform grid than to deal with it in its scattered format. The objective of the modeling step is to apply a mathematical model that best fits the physical model. The model will never be truth. In situations with subsurface problems there is usually only a limited amount of data available because of the high cost of drilling wells to collect new data. The analysts never have as much data as they would like. It's not like topography, where you can go out and stand on the site and see it first hand.

Many phenomena in nature follow a model known as minimum-tension. A computer-generated minimum-tension model can be calculated using an iterative tension reduction method (Briggs 1974). If there is no other information known about the phenomenon except for its value at particular spot locations, then the minimum-tension algorithm provides a smooth, unbiased model of the data. If any additional facts are known about the phenomenon then, of course, that has to be taken into account by applying another model which better fits the situation.

For example, if the phenomenon is moving through a ground water zone and it is known that the zone has an East-to-West flow, then an appropriate flow model should be applied, not a minimum-tension model. Sometimes a flow model will create a non-uniform grid and a minimum-tension algorithm can then be used to model the non-uniform data onto a uniform-grid. This is done when the data needs to be correlated with other non-uniform data sets, or when particular display techniques need to be used that require a uniform grid.

Another modeling technique involves the use of geostatistics to provide the scientist or analyst with information. Geostatistical models, such as kriging, are often used for applications in mining or petroleum exploration. Geostatistical routines are not used to develop a mathematical model of a physical phenomenon like the minimum-tension model does, but rather they strive to develop a block average description of the phenomenon. As the blocks are larger the results seem more valid.

Data editing

The ability to input source data and run a model is very useful, but often the user is confronted with the need to edit the data. Sometimes the source data needs to be queried and/or edited, and other times the model results need to be queried and/or edited. Editing tools, often involving interactive graphic editors, can be applied to 2-dimensional data without a great amount of difficulty, but with 3-dimensional data the problem is much more difficult. Working in 3-dimensions the tools have to be more helpful to the user and have to be graphically more powerful. It is not easy to point and query data locations in 3-dimensions and it is more difficult to edit the source data and then remodel around it.

Three dimensional analysis

One of the more simple techniques that can be used to analyze 3-dimensional data is to apply a set of grid operations. These operations could include such things as:

- grid-to-grid mathematical calculations
- grid refinement
- grid smoothing
- back interpolation
- trend grids

These grid operations would provide a user with a basic set of tools to perform a wide range of analytic functions. Three-dimensional grid models of permeability, porosity, temperature and pressure could be compared, correlated and analyzed together to determine the most likely locations for oil to be found. Hazardous chemical plumes in the ground could be analyzed over time to determine the movement of the plume and any changes in its size or chemical make-up.

When we first think of 3-dimensional problems we often imagine applying analytic tools similar to those we are familiar with in 2-dimensions. This is a very reasonable assumption to start with. However, a 3-dimensional gridded model of a particular phenomenon does not always provide us with the data structure that we need in order to perform some of these operations. One of the solutions to this is to develop 3-dimensional 'iso-surfaces' through the 3-dimensional grids similar to the way we locate 2-dimensional isolines (contours) through 2-dimensional grids. An example of an iso-surface is shown in Plate X.

The iso-surface is a polygonized data structure which is positioned through the 3-dimensional grid where the level of the phenomenon is of equal value. An iso-surface in 3-dimensions is similar to an iso-line or contour line in 2-dimensions. The iso-surface is given its shape by forming small triangular polygons through the gridded data and then connecting these triangles together to form a 3-dimensional surface of equal value, or an iso-surface.

Having the phenomenon defined by user-selected iso-surfaces provides the scientist with an additional set of analytic capabilities and accurate volumes can now be calculated. Iso-surfaces can be intersected by performing 3-dimensional polygon intersection. This operation could allow a scientist to accurately model the movement of a contaminate plume through the ground.

Iso-surfaces can be constrained or limited above and below by 2-dimensional surfaces. This could be used to generate an accurate geologic model where particular materials are limited by geologic structures and faults. Polygons defining features on the earth's surface, such as lease tract boundaries, could be used to cut down through the 3-dimensional iso-surfaces by using a 3-dimensional polygon intersection routine. This would be useful, for example, when a user wants to calculate volumes under lease tracts, or when some underground phenomenon needs to be associated with particular land-use features (Plate XI).

Three dimensional dynamic displays

One of the more powerful and useful functions of a 3-dimensional geoprocessing system is its ability to display information in ways that have never before been seen. This provides the user with a scientific visualization tool that allows him to better understand the phenomenon he is studying and to make more informed decisions about it. The display capability needs to include dynamic movement of the graphic and dynamic selection of options. There is a tremendous advantage in having dynamic capabilities in a 3-dimensional system because of the complexity of the problems. Many of the relationships and features of 3-dimensional phenomena simply cannot be comprehended by the users without these tools.

One of the basic elements of any geographic display is the need to accurately identify the geo-referencing system. The user needs to know where things are, not only in X and Y coordinates, but also in Z coordinates. The user can easily get lost when the system provides the capabilities to rotate the model left, right, up or down. The system needs to provide the user with an appropriate geographic referencing system.

Users need to be able to see and browse through displays of the scattered source point data. They also need to be able to examine and understand the 3-dimensional grid values and how they relate to the source data. Another very informative display of the 3-dimensional model is a colored cube display which presents ranges of the values in the model (Plate XII).

The user needs to be able to slice off edges of the cube display to get views of the inside of the model in order to better understand what the model really looks like (Plates XIII and XIV).

Often the user is searching for a particular value in the data and the system needs to provide tools to locate and display this value. For example, when the ground is contaminated with PCB's as in the sample used here, the user is trying to determine if the values found are above the safe levels for that particular chemical, and if so, what the volume is and where it is located. In a 3-dimensional geoprocessing system the user could select a particular iso-surface level, have the display generated, and then dynamically slice through it to gain a full understanding of the situation (Plate XV).

Phenomena in 3-dimensions are difficult to understand and all of the display functions in a 3-dimensional geoprocessing system need to work together to provide the users with the maximum utility. The user needs to be able to see the source data, to select different iso-surface levels, to assign colors to these levels, to slice edges from the model, to peel off iso-surfaces, to rotate around the display and to zoom in and out. Another useful facility is to see the shape of the iso-surfaces, both above and below a value (Plates XVI and XVII).

Hardware issues

The capabilities needed to work with 3-dimensional data are greatly enhanced by the hardware functions in new graphic workstations. Three-dimensional modeling is compute intensive and is well suited for the new high-performance 3-D graphic workstations. The scientific visualization aspects of a 3-dimensional geoprocessing system are only available by utilizing the 3-dimensional graphic functions in the new workstations. No one view can properly communicate to the user what is going on in a 3-dimensional model. Fortunately for the user community, 3-D workstations are becoming more common in the workplace and prices are dropping.

Conclusion

These new 3-dimensional geoprocessing capabilities are addressing a class of problems that could not be dealt with before. Manual methods can be successful with 2-dimensional problems and many of these are now addressed by computerized systems. The 3-dimensional problems in the earth sciences are generally too complicated to do by hand and computerized systems to date have not been that successful. Applying 2-dimensional tools to 3-dimensional problems has been only moderately successful at best. As the new 3-dimensional geoprocessing tools get into the hands of the users, answers will be discovered to questions that we currently don't understand or even realize we can ask. For earth scientists the move from 2-dimensional geoprocessing into 3-dimensional geoprocessing will be both exciting and rewarding.

References

Briggs, I. C., 1974, Machine contouring using minimum curvature, *Geophysics*, **39** (1) 39-48.

Clarke, K. C., 1986, Advances in Geographic Information Systems, *Computers, Environment and Urban Systems*, **10**, 175-184.

Cowen, D. J., 1988, GIS versus CAD versus DBMS: What Are the Differences?, *Photogrametric Engineering & Remote Sensing*, **54**(11) 1551-1555.

McCormick, B. H., 1987, Visualization in Scientific Computing, *ACSM SIGGRAPH*, **21**(6).

This paper is reproduced, with permission, from the Technical Papers, 1989 ASPRS-ACSM Annual Convention Proceedings, Copyright 1989, by the American Society for Photogrammetry and Remote Sensing and the American Congress on Surveying and Mapping.

Chapter 12

Three dimensional representation in a Geoscientific Resource Management System for the minerals industry

Peter R. G. Bak and Andrew J. B. Mill

Introduction

The extraction of minerals from the earth involves resource evaluation, project design and production planning. Optimisation of these tasks requires the management of both surface and subsurface spatial and textual data and their geoscientific relations. Traditionally, such management has made use of geological and topographical maps, however, with increasing expectations in project definition and control, the two dimensional (2D) methods employed in mapping have become increasingly inefficient and, in many cases, inappropriate. There is a need for development of new techniques which offer enhanced information management capabilities, both in terms of data complexity and volume.

Geographic information systems (GIS), which allow sophisticated resource management of surface features offer an immediate improvement upon traditional management techniques. GIS, however, are limited to modelling 2D and 2.5D entities and, thus, alternative, and complementary, techniques must be considered for the representation of three dimensional (3D) objects. Such representation can be achieved with solid modelling techniques which are commonly used in computer aided design/manufacture (CAD/CAM).

This paper investigates available solid modelling techniques and presents a prototype Geoscientific Resource Management System (GRMS) which incorporates the benefits of boundary representation and volumetric representation, using linear octree encoding. The capabilities of the system are presented and future developments are proposed.

3D representation– an appropriate solid modelling technique

In considering a representation technique for use in resource management, it is important to identify a suitable conceptual model and an appropriate numerical model. The latter determines the computational efficiency with which an entity can be analysed and, thus, determines the functionality of the system and the available application programs. The conceptual model defines the abstract rules on which the numerical model is based and, thus, determines the geometric characteristics of the representation scheme.

Application programs: defining a numerical model

The analytical capability of a solid modelling system is determined by the numerical model employed. The choice of such a model, for a particular system, is dependent on the proposed function of that system. The following capabilities are considered fundamental to geoscientific resource management:

Object definition and modification– the primary function of a GRMS is to facilitate geometric definition, both in terms of object creation and modification. To achieve such definition the system must offer:
- An object definition language;
- Geometric transformations;
- Boolean operations of union, intersection and difference.

Effective visual perception– an effective method of validating the geometry of an object or verifying a design is through cogent visualisation; a GRMS must be capable of:
- 3D orthographic and perspective display;
- Hidden line and surface removal;
- Light source simulation;
- Colour shading;
- View transformation.

The calculation of integral properties– integral properties are essential in both development and production planning; the system must be capable of calculating:
- Volume;
- Surface area;
- Distance;
- Mass (tonnage).

Information retrieval– the ability to represent and manipulate features is of little value unless information associated with each entity can be retrieved. Such retrieval requires:
- Spatial search (spatial query);
- Non-spatial search (relational query).

Characteristics of a representation scheme: the conceptual model

The conceptual model is the manifestation of a set of rules which defines the abstract representation of an entity. Different rules are required for different entities and, therefore, the choice of model is determined by the geometric properties of the objects.

Geoscientific features exist in Euclidean 3D-space. They are *geometrically complete* (Baer, Eastman and Henrion 1979, Mantyla 1983, Sabin 1983) and have the following properties:

Validity– a valid object is:
- Solid– an object is 'solid' if it is defined by a point set which is continuous in 3D-space; it does not contain isolated points or lines;
- Bounded– a bounded object is one which will divide 3D-space into at least two disjoint regions, one of which is infinite. The remaining regions must all be finite;
- Connected– a connected object is one in which there exists a path which connects any pair of points, in the objects point set, without exceeding the bounds of the shape;
- Rigid– a rigid object is one which remains invariant under rigid transformations; local scaling, translation, rotation and reflection.

Spatial uniqueness– an object is spatially unique if the region of space occupied by that object is not coincident with that of any other entity;

Unambiguity– an object is unambiguous if it has a unique topology.

There are three distinct characteristics, common to geoscientific features, which influence the choice of conceptual model:

Structural composition– subsurface features, which are homogeneous[1] can be both solid or fenistrate. For example, a geological structure is 'solid' rock while an underground excavation is 'void';

Geometric complexity– sub-surface features have irregular shapes which can be both convex and concave. Such shapes are difficult to represent using simple mathematical primitives; they are geometrically complex;

Resolution complexity– objects in a universe have different geometric complexities. If the desired accuracy of representation is constant the level of resolution required to model each entity will vary. This variation is referred to as the resolution complexity, which is high for a geoscientific universe.

Having determined the characteristics of the representation scheme and the desired application programs it is possible to examine the available solid modelling techniques and choose which is most appropriate for use in a GRMS.

Representation Techniques

There are six general techniques (Requicha 1982, 1983) commonly used in solid modelling:

1. Wire frame representation– the wire frame technique represents the shape of an object by the set of edges defining the bounding surface;
2. Sweep representation– the sweep technique represents an object by sweeping a defined area or volume along a defined trajectory;
3. Primitive instancing– primitive instancing represents an object by a set of predefined shapes, or mathematical primitives, which are positioned in 3D-space without intersection. An instance of a primitive is defined by a set of numeric values where each value is a parameter in the mathematical equation describing the primitive shape;
4. Constructive solid geometry (CSG)– the constructive solid geometry technique represents an object by combining primitive point sets using Boolean operations (union, intersection and difference);
5. Boundary representation (BR)– the boundary representation technique defines an object by its bounding surface. The latter can be represented as a set of Euler operators (Mantyla 1983) or as a set of coordinates and their connectivity;
6. Spatial occupancy enumeration– spatial occupancy enumeration represents an object by the union of a set of cells where the cell is a primitive shape which can be either regular or irregular. Cells are adjacent, connected and do not intersect.

Geoscientific features are geometrically complete and, thus, the abstract representation of such features must have similar properties. The advantages and disadvantages of the

---

[1] Geoscientific features, when considered as geometric entities, are assumed to be homogeneous bodies bounded by discrete surfaces. When considered as mineral zones, such features are heterogeneous bodies which cannot be defined by discrete values.

aforementioned techniques are discussed with respect to this constraint and the desired application programs.

Advantages and disadvantages of representation techniques

Wire Frame Representation

One advantage of a wire frame technique is an expressive power, which is only dependent on the complexity of the line representations allowed (Mantyla 1983). This characteristic simplifies model generation and offers a large domain[2], which is important for geometrically complex objects.

The most significant disadvantage of the wire frame technique is that object representations are ambiguous, as illustrated in Figure 12.1. Associated with this ambiguity is the inability to generate effective displays, calculate integral properties and uniquely define space.

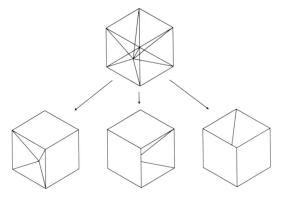

Figure 12.1 An ambiguous wire frame representation

Sweep Representation

There are three types of sweep representation; rotational and translational sweeps, volume sweeps and general sweeps. Of these only the latter, which sweeps a 2D point set about any curve, has a domain which is large enough for use in a geoscientific universe. Unfortunately, the mathematics of general sweeping is unknown (Requicha 1980) and, therefore, integral properties and intersection point sets cannot be calculated in all cases.

Primitive Instancing

Representation using primitive instancing offers a number of advantages:

• Models are easy to generate;
• Integral properties are readily calculated;
• Boolean operations are computationally simple and the resultant point sets may be validated.

These advantages are of value in a geoscientific universe, however, because each primitive is unique, primitive-specific knowledge must be incorporated into the algorithms used to

---

[2]The *domain* of a representation technique is the number of different shapes which can be modelled (Woodwark, 1986).

perform various functions. This limits the number of primitives which can be used and, therefore, the domain of the representation technique.

Constructive Solid Geometry (CSG)

CSG is a commonly used solid modelling technique in CAD/CAM because object creation can be achieved interactively with a simple modelling language. A further advantage lies in Boolean and geometric operations which are computationally simple. If complex primitives are used in a parametric form the modelling domain is large while memory requirements remain small.

Edges and faces in a CSG model are not explicitly defined; it is, therefore, difficult to generate an image from the representation. Generally, CSG based solid modellers overcome this problem by maintaining a *boundary file* which is a BR of the object. This file is created by use of a conversion algorithm and is generated or updated each time a change is made to the CSG model.

The natural method of calculating the integral properties of an object, represented using CSG, is by recursively applying the integral formulas:

$$\int_{A \cup {}^* B} f dV \;=\; \int_A f dV \;+\; \int_B f dV \;-\; \int_{A \cup * B} f\,d\,V \qquad (1)$$

$$\int_{A - {}^* B} f dV \;=\; \int_A \; dV \;-\; \int_{A \cup {}^* B} f\,d\,V \qquad (2)$$

where f is a simple real-valued scalar function, A and B are point sets representing a solid and dV is the volume differential. The starred operators are modified versions of conventional operators and are discussed in the section on 'linear octree manipulation' below.

For a general domain this method can become computationally expensive as the main problem is solving an integral over the intersection of an arbitrary number of primitives (Lee and Requicha 1982).

Boundary representation (BR)

In boundary representation, faces, edges, vertices and their relations are explicitly defined. Such information is essential for graphical display; effective visualisation of complex objects can be achieved using descriptions in this form. For example, modellers based on BR can display images with shading, hidden line and surface removal and light source simulation.

The use of face, edge and vertex primitives offers a large domain, however, calculating the intersection point sets is computationally expensive and reveals a number of disadvantages:

• The validity of the representation is difficult to verify;
• Boolean operations are difficult to perform;
• The calculation of integral properties is a computationally intensive task for complex shapes;
• Complex algorithms are required for model generation.

Spatial occupancy enumeration

Spatial occupancy enumeration represents 3D-space as an array of adjacent cells[3] and is the most literal of the abstract modelling techniques discussed. Such representation, also referred to as *volumetric modelling*, has two distinct advantages:

* Space is uniquely defined– space cannot be occupied by two or more objects; uniqueness of space is assured and need not be verified;
* Cells are spatially indexed– each cell in the array is spatially indexed and is, therefore, addressable; this allows efficient spatial searching.

The domain of this representation technique is, theoretically, infinite. However, memory requirements, which increase with an increase in domain, dictate the number of cells which can be stored. For complex and large domains unlimited resolution of detail cannot be achieved; the representation is, therefore, an approximation of the object.

The edges and vertices defining a cell can be explicitly defined if the cell is a regular shape, such as a cube. An image of the model is generated by displaying each cell in the representation. Such an image is effective but of poor quality.

Choice of representation technique

Investigation into the available solid modelling techniques has indicated that model generation, object visualisation and geometric transformations, for geoscientific features, can be best achieved using boundary representation. Boolean operations, calculation of integral properties and performance of spatial searches are, however, best accomplished using spatial occupancy enumeration.

The prototype GRMS, which is discussed in the following sections, achieves 3D representation of geoscientific features by integrating spatial occupancy enumeration and boundary representation techniques.

A prototype geoscientific resource management system

A prototype GRMS has been developed by the CAD research group, referred to here as 3D-GRMS, which integrates a general purpose computer graphics package, MOVIE.BYU[4],with a volumetric modelling system. The latter is an enhanced version of *Linoct*, a *linear octree* based system developed at the University of Western Ontario (Gargantini 1987).

MOVIE.BYU is a geometric modeller based on BR and used within 3D-GRMS to create 3D representations of geoscientific features and to generate images which offer effective visualisation. The latter is achieved using advanced computer graphics techniques while models are created by *surface triangulation* and an object definition language.

As discussed in the section above, the BR technique is inefficient in calculating integral properties, performing boolean operations and verifying the validity of the model. Further, space is not indexed and, therefore, spatial query is computationally expensive. To overcome these inefficiencies the boundary representation is converted to a volumetric representation–linear octree encoding.

---

[3] The *block model*, commonly used in geoscientific applications, is an example of spatial occupancy enumeration.

[4] MOVIE.BYU is the copyright of the Engineering Computer Graphics Laboratory at Brigham Young University.

Boundary Representation using MOVIE.BYU

The boundary representation technique used in 3D-GRMS is simple in concept. An object is defined by bounding surfaces which are represented by a set of non-intersecting planar polygonal faces; the latter are, themselves, defined by the set of vertices, in each polygon, and their connectivity.

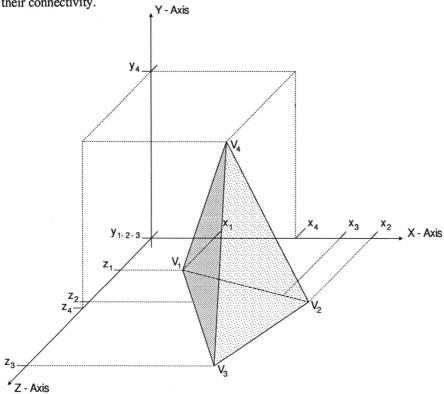

Figure 12.2 Boundary representation of a polyhedron

The data structure consists of a coordinate array and a connectivity array. The coordinate array is the set of coordinates defining the vertices associated with the bounding surface. The connectivity array is the set of vertices defining the planar face polygons where each element in the array is a pointer to the vertex in the coordinate array. For example consider a polyhedron which is bounded by four polygonal faces as shown in Figure 12.2. The boundary representation for this surface would be:

Vertex number	X coordinate	Y coordinate	Z coordinate
1	1256.789	2345.55	423.3
2	1261.938	2345.55	425.1
3	1260.872	2345.55	423.8
4	1259.351	2350.74	427.9
Polygon number	Vertex 1	Vertex 2	Vertex 3
1	1	2	-3
2	1	4	-2
3	1	3	-4
4	2	4	-3

Table 12.1 Representation of a simple polyhedron using a coordinate and connectivity array (the negative sign indicates closure of the polygon).

Model Creation: surface triangulation

Geoscientific features are traditionally represented by boundary loops on 2D maps, plans and sections. Such loops can be digitised, or digitally generated using other techniques, and converted to a 3D boundary representation by means of surface triangulation. The latter is a technique which connects points on 2D planar, parallel contour loops such that the resulting *mesh* can be described by a set of planar polygons. This can be more formally stated as:

Given two borders represented by point sequences p_i, i=0,1,...m-1 and q_j, j=0,1,...n-1 produce a set of triangular faces forming the best spanning surface.

The surface triangulation algorithm used in the 3D-GRMS is based on the shortest span technique developed by Christiansen and Sederberg (Christiansen and Sederberg 1978). The algorithm, which is heuristic in nature, commences by connecting two points, one on each adjacent loop. This line segment is the base of the first polygon. The next line segment is chosen according to which diagonal, between the two loops, has the shortest length, and when connected forms a triangle. This local decision making continues until all the coordinates are meshed together. To illustrate this consider an example: assume two loops, T and B, to be planar and parallel, as shown in Figure 12.3. Initially, points t_1 and b_1 are connected to form the base of the first triangle. The 2 edges of the triangle are created by connecting the *shortest* diagonal between the adjacent loops i.e. $t_1 \rightarrow b_2$. This segment is the base of the next triangle whose edges are also created by connecting the shortest diagonal between the adjacent loops; $t_1 \rightarrow b_3$ or $b_2 \rightarrow t_2$. This process continues until the loops are triangulated as shown in Figure 12.4.

Geological structures are complex and, therefore, the triangulation algorithm does not necessarily generate a geometrically complete shape. A number of features assist in overcoming this problem:

Mapping– if the boundary loops are not mutually centred nor similar in size or shape the resultant triangulation can be wholly incorrect, as illustrated in Figure 12.5. This problem is overcome by translating and scaling the loops onto an arbitrary unit square; the mapping operation centres the loops and reduces them to a similar size;

Branching– a branch is where one or more contour loops exist in the same plane. The triangulation of such loops with a loop in an adjacent plane can be controlled by specifying which coordinate strings must be meshed;

Editing– boundary loops can be interactively edited. The available editing functions are *add*, *delete*, and *move* a coordinate in the boundary loop;

Insertion– if boundary loops, in adjacent planes, are dissimilar in shape an intermediate loop can be inserted between the two planes. The shape of this loop can be edited such that the resultant model is more representative of the object.

The mining project, including an ore body and mine levels, displayed in Plates XVIII and XIX were created by surface triangulation of boundary loops digitised from existing plans.

Model creation– object definition language

Many geoscientific features are difficult to represent as a series of connected 2D boundary loops and, thus, model creation using surface triangulation is not possible. An example of such a feature is a spiral access ramp in an underground mine. To overcome this problem 3D-GRMS is able to generate three general classes of shape using an object definition language. The latter is a program which converts a formal description of a shape into a geometric model using an interactive prompting mechanism.

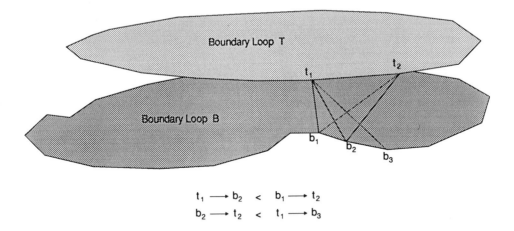

$$t_1 \longrightarrow b_2 \quad < \quad b_1 \longrightarrow t_2$$
$$b_2 \longrightarrow t_2 \quad < \quad t_1 \longrightarrow b_3$$

Figure 12.3 The shortest span method of surface triangulation

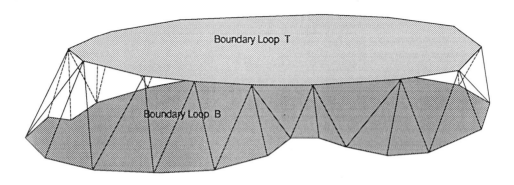

Figure 12.4 Surface triangulation of two boundary loops

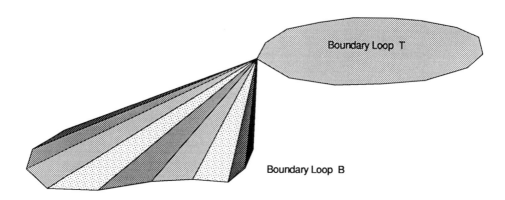

Figure 12.5 Surface triangulation of loops which are not mutually centred

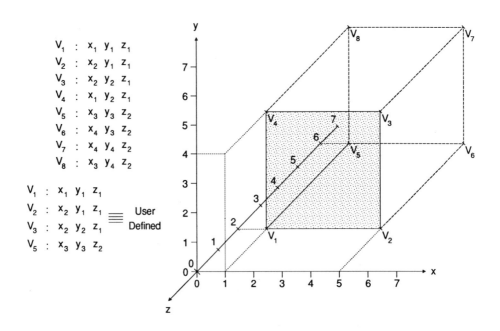

Figure 12.6 Generating a cube using the object description language

The three classes of shape which can be created are:

Hexahedrons– hexahedrons are a class of primitive which can be described by eight corner points and any four cornered quadrilateral surface. Hexahedrons with a constant prismatic cross-section are described by simply entering the coordinates of the vertices of the object. Non-prismatic hexahedrons require more information detailing the shape of quadrilateral surfaces (Christiansen and Stevenson 1986). For example, to create a parallelpiped the user is prompted for the coordinates of four vertices; three on the front face and one on the back face. This is illustrated in Figure 12.6;

Ellipsoids– this class of primitive includes ellipsoid surfaces and thick-shelled volumes whose boundaries are ellipsoids. Such primitives can be described by the radii of the outer and inner ellipsoids, the centre of the ellipsoid and a vector which sweeps through a latitude and a longitude (Christiansen and Stevenson 1986), as illustrated in Figure 12.7. An example of a thick-shelled volume is a torus;

Volumes of revolution– this class of primitive includes surfaces and volumes of revolution, spirals of translation and radial spirals. Such primitives can be described by a point set defining a surface or volume, an angle of revolution and, where required, a translation along the axis of revolution (Christiansen and Stevenson 1986) as illustrated in Figure 12.8.

The models generated using the object definition language have an arbitrary location in space and, therefore, geometric transformations must be performed to position an object within the model universe. To illustrate this capability consider the shortest vertical shaft displayed in Plate XIX. This object was created by revolving a surface, described by four coordinates, about the X-axis. The radius of curvature is the radius of the shaft while the length of the planar surface is the depth of the shaft. To position the shaft in the mining project universe three transformations were performed; a translation of the origin, a rotation of the X-axis and a rotation of the X-Y plane.

Model Display

The display capabilities of the 3D-GRMS are extensive including; light source simulation, shadow casting, transparency, shading, fogging, hidden line and surface removal, world window manipulation and view control. These capabilities, which allow the user to visually *roam* around any feature, are essential in graphically verifying an object and determining whether the representation is an adequate 'visual likeness' of the desired model.

It is beyond the scope of this paper to discuss the techniques used in image generation, however, Plates XVIII-XX provide an indication of the graphical power of the system[5]. The number of polygons displayed and the time taken to generate each image is given in Table 12.2.

Image	No. of polygons	Display time (secs.)
Plate XVIII	498	47.37
Plate XIX	2163	107.12
Plate XX	2741	151.20

Table 12.2 Display Times for Plates XVIII-XX

---

[5] 3D-GRMS is operational on a Sun 360 Colour Graphics Workstation, running Sun OS 4 with 4Mb of main memory and a floating point processor.

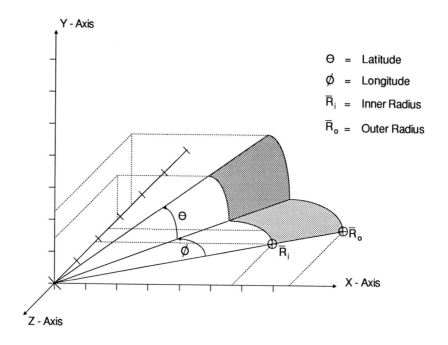

Figure 12.7 Generating ellipsoids using the object description language

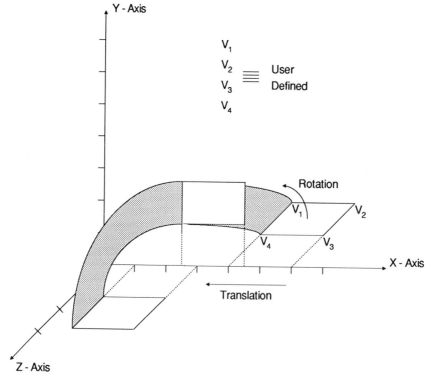

Figure 12.8 Generating volumes of revolution using the object description language

Volumetric Representation using Linear Octree Encoding

The linear octree encoding algorithms in 3D-GRMS comprises three processes:

- *Octree generation–* boundary representation to Octree conversion and 'Brute Force' generation;
- *Octree display–* colour display and view transformations;
- *Octree manipulation–* boolean operations, calculation of integral properties and information retrieval.

The linear octree data structure is fundamental to the encoding of these algorithms.

The linear octree data structure

Assume a finite region of space to exist in a cuboid, referred to as the universe. The cuboid is decomposed into eight sub-cubes, of equal size, called octants, which are indexed by an encoding scheme. Each octant is attributed a label or colour depending on whether it is inside (black), outside (white) or partially within (grey) the finite region of space (Meagher 1982). Grey octants are recursively decomposed until there are no more grey octants or a limit or recursion has been attained. In the latter case an octant is referred to as a *volume pixel* or voxel (Mark and Cebrian 1986). The resultant universe can be visualised as a 3D raster of voxels or block model with varying size blocks, an example of which is given in Figure 12.9.

The abstract representation of the modeled region of space is a tree structure whose root corresponds to the whole universe. Each grey octant is the *parent* node of eight suboctants where each suboctant may, itself, be a parent node or a *leaf* node. The latter can be either a black octant, a white octant or a voxel. The complete tree structure is called an *octree* and can be simulated in a computer by means of the encoding scheme and a set of *pointers*, which link parent nodes with antecedent nodes (Meagher 1982). Figure 12.10 is the octree representation of the region of space in Figure 12.9.

The *linear octree* is a compressed form of the regular octree described, in which only the black leaf nodes are stored (Gargantini 1982). The label used to index each octant is a unique key which identifies the path from the root of the tree to that octant (leaf node) and, therefore, dispenses with the need for pointers. Each key is a string of R octal digits where R is the maximum number of decompositions or the resolution of the octree universe.

The value of an octal digit identifies the position of an octant in an integer coordinate system according to the notation given in Figure 12.11 (Gargantini 1982). This value can be calculated from the 3D world coordinate of a known vertex of the octant using the encoding techniques described by Gargantini (Gargantini 1982). The position of the digit in the string defines the number of decompositions that have occurred i.e. the *level of resolution*. This indexing scheme is illustrated by the example in Figures 12 and 13.

To improve the 'functionality' of the linear octree a number of attributes are associated with each key. These are:

Grouping Factor– when a leaf node is not a voxel the key for the black octant will have less than R octal digits and this complicates the simple mode of computer storage. Each key is, therefore, padded out with a string of 0s until there are R digits. The grouping factor defines the number of 0's added to the key and is an integer 0,1,.,R. Thus, a voxel will have a grouping factor of 0 and an octant, at a resolution of 2, will have a grouping factor of R-2.

Region Number– within the universe there may exist a number of discrete objects. To identify the leaf nodes which define an object, a region number, or colour value, is assigned to each key.

Figure 12.9 Universe recursively decomposed into octants and voxels

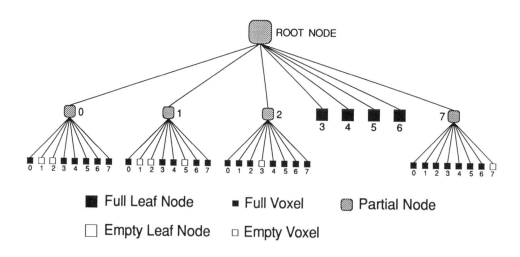

Figure 12.10 Abstract representation of an octree structure

Surface Bit Array– the surface bit array, which has six elements, stores information describing the condition of each face associated with the octant indexed by the key. If the face of an octant is adjacent to an octant with a similar region number, that face is labeled as unblocked (0). If there is no adjacent octant i.e. the octant forms part of an exposed surface, or the adjacent octant has a different region number, the face is labeled blocked (1) and assumed to exist on the surface of the modeled entity.

Once the keys have been assigned a grouping factor, region number and surface bit array they are sorted in ascending order, as shown in Table 12.3 which is the full linear octree for the universe in Figure 12.12.

Key	Grouping	Region	Surface Array					
Octal	Factor	No.	E	S	B	W	N	F
002	0	2	0	0	0	1	1	1
003	0	2	0	0	0	0	1	1
004	0	2	0	0	0	1	1	1
005	0	2	1	0	0	0	1	1
006	0	2	0	1	0	1	0	0
007	0	2	1	1	0	0	0	0
012	0	2	1	0	1	0	1	1
020	0	2	0	0	1	1	0	1
021	0	2	0	1	1	0	0	1
022	0	2	1	1	1	1	0	1
030	0	2	0	1	1	0	0	1
031	0	2	1	0	1	0	1	1
033	0	2	1	0	1	1	0	1
040	1	2	1	1	0	1	1	0
211	0	3	0	1	1	1	0	1
300	0	4	0	1	1	0	1	1
301	0	4	0	1	1	0	1	1
310	1	4	1	1	1	1	1	1
351	0	4	1	1	0	1	1	0
355	0	4	1	1	0	1	1	0
400	1	1	1	1	0	1	1	0
440	1	1	1	1	1	1	1	0
454	0	1	0	0	1	0	1	1
455	0	1	1	0	1	0	1	1
456	0	1	0	0	1	0	0	1
457	0	1	1	0	1	0	0	1
474	0	1	0	0	1	1	0	1
475	0	1	1	0	1	0	0	1
476	0	1	0	1	1	1	0	1
477	0	1	1	0	1	0	0	1
655	0	3	0	1	1	1	0	1
700	2	6	1	1	1	1	1	1

Table 12.3 Complete linear octree for Figure 12.12.

Linear Octree Generation

A linear octree can be generated by using either the Double Connectivity Filling (DCF) algorithm (Chan, Gargantini and Walsh 1986) or the 'Brute Force' algorithm.

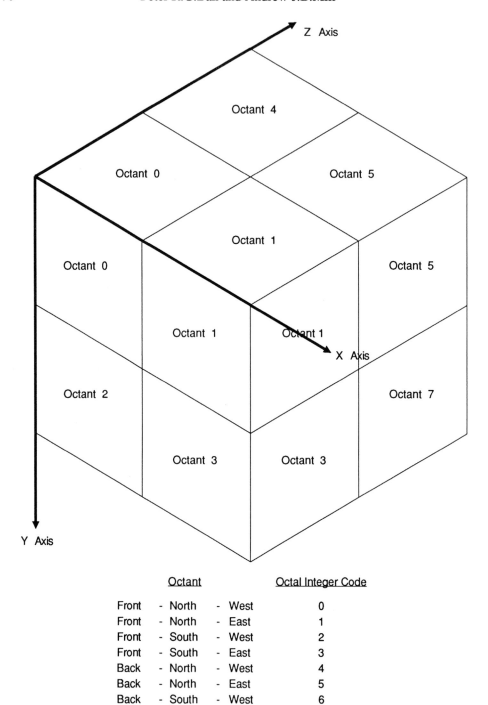

Octant			Octal Integer Code
Front	- North	- West	0
Front	- North	- East	1
Front	- South	- West	2
Front	- South	- East	3
Back	- North	- West	4
Back	- North	- East	5
Back	- South	- West	6
Back	- South	- East	7

Figure 12.11 Octal notation for an integer coordinate system

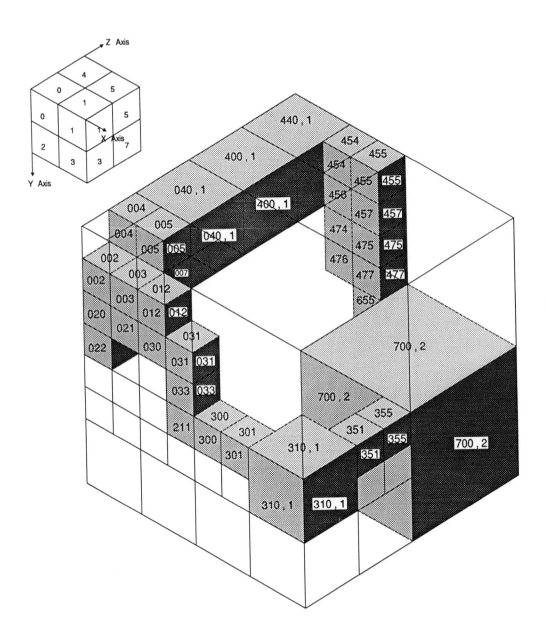

Figure 12.12 A complex object decomposed into octants and voxels

The *DCF algorithm* converts the boundary representation of a region of space into a linear octree by a label propagation technique (Tamminen and Samet 1984) and octant aggregation (Atkinson, Gargantini and Walsh 1986). It is assumed that each polygon, defining the boundary surface of the object, is planar. Each edge of the polygon is projected onto a plane which lies parallel to one of the principle axes. The edges of the projected polygon are converted to voxels using an approach similar to Bresenhams Line algorithm (Foley and van Dam 1982). With each edge represented by voxels, and by assuming all voxels must be 'face' adjacent[6], the projected polygon can be 'filled'. Once the projected polygon has been converted, the set of voxels are back-projected to the original location of the polygon.

The surface bit arrays of the voxels representing each polygonal face of the surface boundary are defined according to the normals of each polygonal face. These bit arrays identify the interior and exterior of the object and control the adjacency search which 'fills' the 'surface' octree, towards its centre, with octants (octant aggregation). To ensure that the octree conversion does not leave any 'holes' in the object octree connectivity labeling is used. This technique applies the principal of separating maximally-connected subsets, which determine the interior of a set of 3D regions, by letting a part of the universe border (which is by definition external) expand towards each region. In this fashion the regions interior is encapsulated by the surrounding empty universe and, thus, the desired octree is the inverse of the generated octree.

The BR models shown in Plates XVIII and XIX were converted to linear octrees using the DCF algorithm. The conversion times and the number of nodes in each octree are shown in Table 12.4.

Element	Conversion time (secs)	Number of Polygons	Number of Nodes
Level 1	6.7	182	8200
Level 2	10.6	254	13870
Level 3	17.5	318	22420
Level 4	20.4	340	26655
Level 5	18.2	289	23517
Level 6	16.5	282	21220
Level 7	12.4	242	15625
Level 8	7.5	148	9900
Shafts	20.1	108	21392
Low grade zone	293.0	720	422662
High grade zone	89.1	498	120943
Open pit	113.8	80	68503

Table 12.4 Conversion times from boundary representations to linear octree representation.

The *brute force algorithm* is designed to generate linear octrees for shapes which can be defined by mathematical equations i.e. 'primitive' objects. Initially a universe is decomposed into a 3D raster of $2^R * 2^R * 2^R$ voxels. A point membership test is performed to determine whether a voxel lies within the boundary of the primitive object, and labeled black or white. After the entire raster has been inspected the black voxels are *condensed*, where possible, into octants (Gargantini 1982). An *adjacency search* algorithm is used to determine whether an octant has a neighbouring octant, in any of the six principle directions (E,W,N,S,B,F), and labels the surface bit array accordingly (blocked or unblocked) (Gargantini 1982).

---

[6] Face adjacency assumes octants to be connected via their faces rather than by edges or vertices.

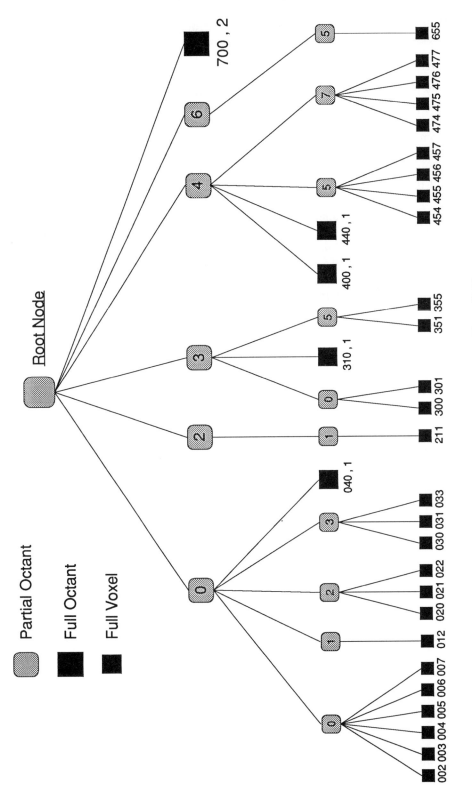

Figure 12.13 The abstract representation of the object in Figure 12.12

The brute force algorithm is, at present, capable of generating the linear octree for a sphere, an ellipsoid, a slab and a cylinder.

Linear Octree Display

The display of a linear octree is based on a back-to-front algorithm which produces an axonometric projection of each octant (Gargantini, Walsh and Wu 1986). Visual realism is improved by use of depth shading, the presence of a light source and propagation of the display from the back to the front. The latter is a 'hidden surface removal' technique in which objects are displayed in order of their position with respect to the view point.

The algorithm proceeds by defining a direction of view using a 3D transformation, sorting the keys of the linear octree in a 'back-to-front' order and 'painting' the octants on to the screen.

The octree universe is assumed to exist in a right handed world coordinate system, with axes $\{x,y,z\}$. The direction from the origin of this coordinate system to the viewer is given by a vector $\{x_v,y_v,z_v\}$, called the view plane normal (VPN). The display screen is assumed to be normal to the VPN at all times and, thus, a second coordinate system exists, referred to as the view reference coordinate system (VRC), with axes $\{U,V,N\}$ where N coincides with the VPN (Gargantini, Walsh and Wu 1986). Figure 12.14 illustrates the relation between these coordinate systems.

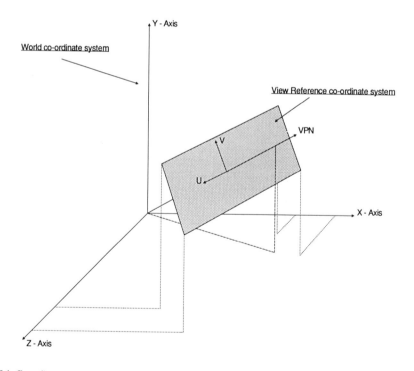

Figure 12.14 Coordinate systems used in linear octree display

To view the object from a chosen direction, defined by the VPN, the $\{x,y,z\}$ axes must be mapped on to the $\{U,V,N\}$ axes. This is achieved in two steps; firstly rotating the x and z axes about the y axis, such that the positive z axis coincides with the projection of the VPN on the x–z plane; and secondly rotating the y and z axes about the x axis such that the positive z axis coincides with the VPN. The x axis maps on to the U axis and the z axis maps on to the VPN (Gargantini, Walsh and Wu 1986)

The rotations are performed about the origin of the coordinate system which is conventionally at (0,0,0). Such rotations position the object outside the defined universe (3D raster) and, thus, a translation is required to re-position the object within the field of view. The complete transformation is given by a matrix called the *view matrix* (Gargantini, Walsh and Wu 1986):

$$\text{view matrix} = \begin{pmatrix} \sin\theta & -(\cos\theta)(\sin\phi) & (\cos\theta)(\cos\phi) & 0 \\ 0 & \cos\phi & \sin\phi & 0 \\ -\cos\theta & -(\sin\theta)(\sin\phi) & (\sin\theta)(\cos\phi) & 0 \\ 2^{(R-1)}f(\theta,\phi) & 2^{(R-1)}g(\theta,\phi) & 2^{(R-1)}h(\theta,\phi) & 1 \end{pmatrix} \quad (3)$$

where

$$
\begin{aligned}
f(\theta,\phi) &= 1-\sin\theta + \cos\theta \\
g(\theta,\phi) &= 1-\cos\phi + (\cos\theta + \sin\phi)(\sin\phi) \\
h(\theta,\phi) &= 1-\sin\phi + (\cos\theta + \sin\phi)(\cos\phi) \\
\sin\theta &= \frac{z_v}{D_v}, \cos\theta = \frac{x_v}{D_v}, (0 \le \theta \le 2\pi) \\
D_v &= (x_v^2 + z_v^2)^{\frac{1}{2}} \\
\sin\phi &= \frac{y_v}{D_v}, \cos\phi = \frac{D_v}{D}, (-\frac{\pi}{2} \le \phi \le \frac{\pi}{2}) \\
D &= (x_v^2 + y_v^2 + z_v^2)^{\frac{1}{2}} .
\end{aligned}
$$

Hidden surface removal is achieved by displaying octants in order of their distance from the viewer, with the furthest being displayed first i.e. a back-to-front order. As the position of the viewer changes the distance to each octant changes and, thus, the order of display varies for a given view direction. There are 14 such orders of priority, or priority orders; 6 central projections and 8 discrete regions of view in which the order of display does not vary. Table 12.5 lists the possible priority orders, while Figure 12.15 illustrates the ordering for one discrete region of view.

One characteristic of an octant is that, at most, only three adjacent faces are visible at any one time and, therefore, if the direction of view is known i.e. the VPN, it is possible to determine the visibility of each face, as shown in Table 12.6. This characteristic can be used to reduce the number of faces displayed and, thus, reduces the required computation.

To illustrate the key sorting process consider the finite region of space in Figure 12.12. If the VPN has coordinates:

$$x_v<0, y_v>0, z_v>0 \quad (4)$$

the priority order is 1,0,3,5,2,4,7,6. There are no nodes with a 1 as the first octal digit and, therefore, all nodes with a 0 as the first octal digit are considered first. The priority order is recursively applied such that the set of nodes with a first octal digit of 0 are sorted into a back-to-front order. This process continues until the linear octree has been completely sorted. The final list for the example is:

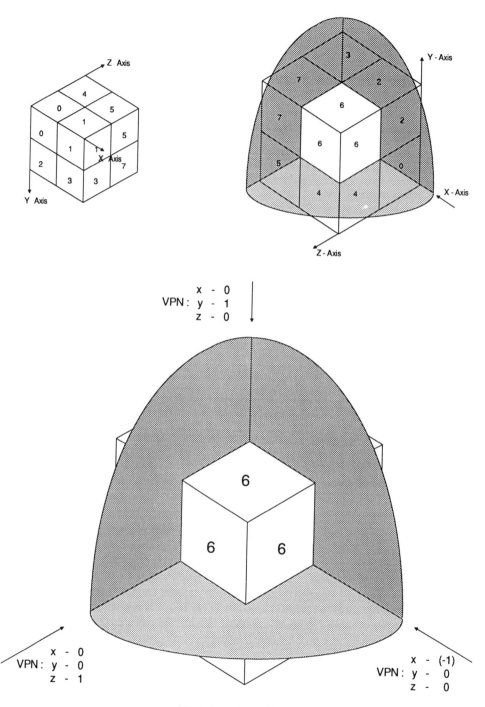

VPN : (x < 0   y > 0   z > 0)

Priority Order : { 1,0,3,5,4,2,7,6 }

Figure 12.15 Priority Order for a known VPN

VPN			Priority Order
x_v	y_v	z_v	
<0	<0	<0	7,6,5,4,3,2,1,0
>0	<0	<0	6,7,4,2,5,3,0,1
<0	>0	<0	5,4,7,1,6,0,3,2
>0	>0	<0	4,5,6,0,7,1,2,3
<0	<0	>0	3,2,1,7,0,6,5,4
>0	<0	>0	2,3,0,6,1,7,4,5
<0	>0	>0	1,0,3,5,2,4,7,6
>0	>0	>0	0,1,2,3,4,5,6,7
Central Projections			
=1	=0	=0	4,0,6,2,5,1,7,3
=(-1)	=0	=0	3,7,1,5,2,6,0,4
=0	=1	=0	1,0,5,4,3,2,7,6
=0	=(-1)	=0	6,7,2,3,4,5,0,1
=0	=0	=1	0,1,2,3,4,5,6,7
=0	=0	=(-1)	7,6,5,4,3,2,1,0

Table 12.5 Priority Order for Octant Labeling according to VPN Coordinates.

VPN	Visible Faces
$x_v < 0$	Western
$x_v > 0$	Eastern
$y_v < 0$	Southern
$y_v > 0$	Northern
$z_v < 0$	Back
$z_v > 0$	Front

Table 12.6 Visibility of an octant with respect to VPN coordinates.

$$\{012,003,005,002,004,007,006,031,030,033,021,020,$$
$$022,040,310,301,300,351,355,211,400,455,454, \qquad (5)$$
$$457,456,440,475,474,477,476,700,655\}$$

Once the linear octree has been sorted into a back-to-front order the surface bit array, associated with each key, is investigated to determine whether any faces are blocked i.e. exposed to the background. If a blocked face is not identified the key is discarded as the octant does not lie at the surface of the object.

The visual representation of an octant is the projection of a face or faces on to the display screen. Each key, therefore, must be converted into a set of screen coordinates defining the associated projection. This is achieved by:

1. Decoding a key into the corresponding world coordinate;
2. Mapping the world coordinate into the VRC system by applying the *view matrix*;
3. Calculating the VRC of the vertices, required to define the face or faces of the octant, using the VPN, the grouping factor and the VRC of the world coordinate;
4. Mapping the VRC for the set of vertices into the screen coordinate system.

Once the screen coordinates of an octant have been calculated they are displayed as a polygon and *fast* shaded. The latter is a simple form of shading where the polygon is filled with a single colour shade which is determined by depth and the effect of a light source (Atkinson, Gargantini and Wu 1987). Plates XXI and XXII show the linear octree display of the mining project represented using BR models in Plates XVIII–XX. The display times for these octree models, which contain 846409 octants, is in the order of 3800 seconds.

Linear Octree Manipulation

Linear octree manipulation algorithms include boolean operations, the calculation of integral properties and information retrieval.

Conventional boolean operations, when applied to solid modelling, do not necessarily generate geometrically complete representations. To overcome this problem solid modelling theory uses *regularised* set operations, denoted by $\cap^*, \cup^*, -^*$, which are modified versions of their classical counterparts (Requicha 1980). In octree theory these regularised set operations need to be further modified.

A linear octree represents a unique region of space and, by use of region numbers, defines the occupancy of space. Thus, when a set operation is performed on two octrees both the *occupation* and *composition* of space must be considered. The octree set operations are defined using symbolic writing:

Octree join– a classical union of two point sets A and B is defined by:

$$A \cup B = C = \{x | x \in A \text{ and/or } x \in B\} \tag{6}$$

The octree equivalent is defined by

$$A ><_o^* B = C = \{x | x \in A \text{ and } x \in \{B\text{-}A\}\} \tag{7}$$

where the set C is complete. This operation is termed a join ($><_o^*$) because the composition of the set $\{x | x \in \{A \cap B\}\}$ is that of the set $\sim \{x | x \in \{A\text{-}B\}\}$.

Octree union– the octree union operation ($\cup_o^*$) is a special case of a join where the composition of the resultant set C is homogeneous; set B is an *extension* of set A;

Octree intersection– the octree intersection operation ($\cap_o^*$) is defined by

$$A \cap_o^* B = C = \{x | x \in A \text{ and } x | x \in B\} \tag{8}$$

where C is complete and has the composition of the set $\sim\{x | x \in \{A\text{-}B\}\}$.

Octree difference– an octree difference operation ($-_o^*$) is defined by

$$A -_o^* B = C = \{x | x \in A \text{ and } x | x \notin B\} \tag{9}$$

where C is complete and has the composition of A.

A linear octree is a sorted list of integer keys and, therefore, an octree set operation is achieved by a tree traversal and a comparison test. Such computations do not require floating point processing and are recursive in nature. Hence, they are computationally efficient. Table 12.7 lists the computation times required for the octree boolean operations used to create the mining project model displayed in Plates XXI and XXII. (A Op_o^* B = C).

Octree Set (A)	OctreeSet (B)	Operation	Nodes (A)	Nodes (B)	Nodes (C)	Time (secs)
Level-1	Level 2	join	8200	13870	22070	3.1
Level 1	Level 2	join	8200	13870	22070	3.1
Level 3	Level 4	join	22420	26655	49075	7.5
Level 5	Level 6	join	23517	21220	44737	6.6
Level 7	Level 8	join	15625	9900	25525	3.7
Level 1-2	Level 3-4	join	22070	49075	71145	10.8
Level 5-6	Level 7-8	join	44737	25525	70262	10.5
Level 1-4	Level 5-8	join	71145	70262	141407	21.2
Low-grade	High-grade	join	422662	121022	533558	79.6
Shafts	Level 1-8	join	21392	141407	162514	24.3
Open pit	Mine tunnels	join	168503	162514	331017	49.6
Mine	Ore zone	join	331017	533558	846409	122.1
Mine	Ore zone	intersect	331017	533558	18166	70.7
Mine	Ore zone	minus	331017	533558	312851	96.7

Table 12.7 Computation times for Boolean operations used to create the example mining project.

The integral properties which are of interest are volume, surface area, distance and mass. If a linear octree is homogeneous i.e. each node has a common region number, these properties can be calculated by using the following equations:

$$V = \left(\sum_{g_0=0}^{R} f_v\,(g_0) \times 8^{g_0} \right) \times K_v \qquad (10)$$

$$A = \left(\sum_{g_0=0}^{R} f_a\,(g_0) \times 2^{g_0} \right) \times K_a \qquad (11)$$

$$D = ((x_2 - x_1)^2 + (y_2 - y_1)^2 + (z_2 - z_1)^2)^{\frac{1}{2}} \qquad (12)$$

$$M = \left(\sum_{g_0=0}^{R} f_\rho\,(g_0) \right) \times K_v \qquad (13)$$

where

V	=	volume
A	=	surface area
D	=	distance
M	=	mass
R	=	resolution of the octree universe
x_n, y_n, z_n	=	world coordinates of 'front-north-west' vertex of octant n
g_0	=	grouping factor
$f_v(g_0)$	=	$\displaystyle\sum_{g_0}$ OCTANT
$f_a(g_0)$	=	$\displaystyle\sum_{g_0}$ BORDER FACES

$$f_\rho(g_o) \quad = \quad \sum_{g_o} OCTANT \times \rho_o \times 8^{g_o}$$

$$K_v \quad = \quad (\text{scale factor})^3$$

$$K_a \quad = \quad (\text{scale factor})^2$$

$\rho_o \quad = \quad$ Density of an octant

OCTANT $\quad = \quad$ Octant with a grouping factor of g_o

BORDER FACES $\quad = \quad$ Visible faces of a border octant with a grouping factor of g_o

These equations are recursive and, therefore, the calculation of integral properties, for homogeneous linear octrees, is computationally simple. If the linear octree is not homogeneous a *connected component* labeling algorithm must be used to distinguish between region numbers.

The connected component labeling algorithm, developed by Gargantini (Gargantini & Tabakman 1983) and implemented in the 3D-GRMS, investigates whether an octant has a neighbour and determines its connection status. Octants can be connected in three ways:

1. Face Connected– if an octant has a neighbour with a common face it is face connected with that neighbour;
2. Edge Connected– if an octant has a neighbour with a common edge, and is not face connected, it is edge connected with that neighbour;
3. Vertex Connected– if an octant has a neighbour with a common vertex, and is neither face connected nor edge connected, it is vertex connected with that neighbour.

Neighbouring octants are assumed to belong to a common point set (object) if they are face connected and have similar region numbers. Such octants are *labeled* by means of an adjacency search. This procedure continues until the entire space occupied by the linear octree has been labeled into constituent components. The integral properties of each entity represented by the linear octree can, thus, be calculated using the aforementioned equations.

Information Retrieval

The information retrieval capabilities of the 3D-GRMS include spatial search, non-spatial search and an interactive facility referred to as *tracking*.

The hierarchical nature of the linear octree data structure is ideally suited to spatial search. Each key explicitly defines the location of an octant or voxel in the universe and with respect to the root of the abstract tree structure. It is possible, therefore, to calculate the key of any neighbouring octant whether it be an adjacent neighbour or a distant neighbour i.e. at a location defined by a distance and a direction (vector). The adjacency algorithm previously discussed is based on this capability.

There are three distinct types of spatial search which can be performed:

Region Search– a region search is one in which all octants and voxels, in a search window, are investigated to identify the existence of a key attribute e.g. region number. The window can be defined by an arbitrary boundary, a radius of influence or a linear octree;

Border Search– the border search is a region search where the key attribute is the surface bit array and the 'window' is a linear octree which represents a single entity;

Connected Search– a connected search is one in which the search window is defined by the connectivity of octants or voxels with a given attribute.

The linear octree is a sorted list of integer octal keys and, therefore, the spatial search functions utilise binary search and *bitwise operators* (Kernighan and Ritchie 1978).

Non-spatial data are the attributes associated with each key in a linear octree. The retrieval of such data is achieved by a *sequential* search of the list of keys. For example, to identify all octants in a linear octree, which have a grouping factor of 2 and a region number of 3, each key in the list must be examined in sequence.

Tracking is a utility which calculates the world coordinates of any octant or voxel which is visible on the display screen. This is achieved by reversing the display process: the x and y coordinate values are calculated from the screen coordinates while the z value is determined by the colour intensity i.e. depth shading. A light pen or 'mouse' is used to identify the desired point in the image and, thus, the tracking facility can be used to digitise a region of space which is of interest or interactively interrogate any entity.

Discussion

Examination of the development of data structures reveals that innovation has been, in many cases, fueled by a need to extend the limits imposed by hardware upon software capability. Rapid changes is machine architecture and improvements in hardware specifications not uncommonly render the advantages of software solutions ephemeral and the techniques outmoded. This has been true for spatial occupancy enumeration which, as illustrated in this paper, offers no special benefits when considered as a representation technique for use in classical CAD applications; object creation is complex, 3D visualisation is inefficient and a degree of inaccuracy is encountered when calculating integral properties. The use of linear octree encoding may, therefore, seem inappropriate considering the recent development of 'super-computer' graphics workstations[7] which are capable of storing, manipulating and displaying BR models in real time.

The emergence of both 3D GIS and GRMS and the growing interest in the use of databases of all forms, focuses attention upon the importance of versatile spatial indexing and spatial query through graphical user-interfaces. The design of such interfaces will exploit new advances in 3D visualisation, however, the inclusion of animation and analysis of the dynamics of engineering and scientific systems, as a capability in CAD, will further emphasis the need for efficient indexing of events and features in a temporal and spatial context. It is for this reason that the integration of boundary representation and spatial indexing techniques, such as linear octree encoding, will find wide and enduring application in the geoscience disciplines.

The 3D-GRMS prototype system described in this paper embraces new and advanced techniques merging the conventional bounds between CAD and information technology to produce a prototype for future spatial analysis systems.

Acknowledgements

This project has been funded by the Science and Engineering Research Council and James Capel and Co. Limited. The authors wish to thank Dr. Irene Gargantini and Harvey Atkinson for their technical contribution, Professor Tosiyasu Kunii for his interest and advice and many other individuals for their support.

---

[7] Stellar, Ardent and Alliant are examples of super-computer graphics workstations which are capable of rendering 50,000 to 100,000 Gouraud shaded polygons per second and manipulating upto 600,000 3D vectors per second.

References

Atkinson, H. H., Gargantini, I. and Walsh, T. R., 1986, Filling by quadrants and octants. *Computer Vision Graphics and Image Processing*, **33**, 138–155.
Atkinson, H. H, Gargantini, I. and Wu, O. L., 1987, LINOCT 2.0. *Technical Report* **176**, The University of Western Ontario, Department of Computer Science.
Baer, A., Eastman, C. and Henrion, M., 1979, Geometric modelling: a survey. *Computer Aided Design*, **11**(5), 253–272.
Chan, K. C., Gargantini, I. and Walsh, T. R, 1986, Double connectivity filling for 3D modelling. *Technical Report* **155**, The University of Western Ontario, Department of Computer Science.
Christiansen, H. N. and Sederberg, T. W., 1978, Conversion of complex contour line definitions into polygonal element mosaics. *ACM Computer Graphics*, **12**(3), 187–192.
Christiansen, H. N. and Stevenson, M., 1986, MOVIE.BYU Training text. Engineering and Computer Graphics Laboratory, Provo, Utah.
Foley, J. D. and van Dam, A., 1982, *Fundamentals of interactive computer graphics*. (Reading, MA: Addison-Wesley).
Gargantini, I., 1982, Linear octrees for fast processing of three-dimensional objects. *Computer Graphics and Image Processing*, **20**, 365–374.
Gargantini, I. 1987 Personal Communication.
Gargantini, I. and Tabakman, Z., 1983, Separation of connected components using linear quad- and octrees. *Congressus Numerantium*, **37**, 257–276.
Gargantini, I., Walsh T. R. and Wu, O. L., 1986, Viewing transformations of voxel-based objects via linear octrees. *IEEE Computer Graphics and Applications*, **6**(10), 12–21.
Kernighan, B. W. and Ritchie, D. M., 1978, *The C Programming Language*. (Englewood Cliffs, New Jersey: Prentice-Hall, Inc.)
Lee, Y. T. and Requicha, A. G., 1982, Algorithms for computing the volume and other integral properties of solids. I. Known methods and open issues. *Communications of the ACM*, **25**(9), 635–641.
Mantyla, M., 1983, Solid modelling: theory and applications. *Eurographics Tutorials '83*, pp. 391–425.
Mark, D. M. and Cebrian, J. A., 1986, Octrees: a useful data-structure for the processing of topographic and sub-surface area. *Proceedings ACSM-ASPRS Annual Convention 1986, 'Cartography and Education'*, pp. 104–113, Washington, D.C., March 1986.
Meagher, D., 1982, Geometric modeling using octree encoding. *Computer Graphics and Image Processing*, **19**(2), 129–147.
Requicha, A. A. G., 1980, Representations for rigid solids: theory, methods, and systems. *ACM Computing Surveys*, **12**(4), 437–464.
Requicha, A. A. G. and Völcker, H.B., 1982, Solid modelling: a historical summary and contemporary assessment. *IEEE Computer Graphics and Applications*, **12**(2), 9–24.
Requicha, A. A. G. and Völcker, H.B., 1983, Solid modelling: current status and research directions. *IEEE Computer Graphics and Applications*, **13**(7), 25–37.
Sabin, M. A., 1983, Geometric modelling– fundamentals. *Eurographics Tutorials '83*, pp. 343–389.
Tamminen, M. and Samet, H., 1984, Efficient octree conversion by connectivity labeling. *ACM Computer Graphics*, **18**(3), 43–51.
Woodwark, J., 1986, *Computing Shape*. Sevenoaks,England: Butterworths.

Index

1 dimensional 13
2 dimensional 1-2
2.5 dimensional 17
3 dimensional 17
3D digitiser 17, 181

Accomodation 105-6
ADA 110
aeromagnetic survey 138 et seq.
aeronautic charts 92
Aleutian Islands 6
alluvial deposits 41, 49, 120
animation 80
anti-aliasing 89
Appalachian, north, orogen 137-8
aquifers 12, 117 et seq.
ARC/INFO 139
architecture 99
aspect 54 et seq., 85
atmospheric attenuation 91
attribute geometry 15, 54
attribute topology 54
Autocad 110
axes 87, 106, 172
axonometric projection 87

Binary search 180
Bingen 64
blending function 22, 28, 32
block diagrams (see also isometric and
    perspective view models) 1, 49, 146
Boolean operations 16, 156
borehole 4, 120
Bouguer gravity data 138 et seq.
boundary representation (BR) 157, 159,
    172, 177, 181
BR 157, 159, 172, 177, 181
brain 2, 87
break lines 21
Bresenhams line algorithm 172

Bridge-Leeder model 120
Brownian motion 83
brute force algorithm 169, 172

C 110
CAD 18, 80, 94, 110, 115, 117, 126,
    155, 159, 181
cartography 100
cathode ray tube 1, 84
CATIA 110
Cauchy-Schwartz inequality 44
CAVIA 93
channel 132
choropleth maps 37, 109
chromostereopsis 104, 106
closed depressions 60
CMP 140
co-kriging 13, 37 et seq., 120
coasts 13
colluvial deposits 41
colour models 85
colour shading 105
common mid point (CMP) 140
computer aided design (CAD) 18, 80, 94,
    110, 115, 117, 126, 155, 159, 181
computer hardware 8, 18, 80, 153
conceptual design 11 et seq., 58, 116,
    122, 155
connected component labeling 180
constructive solid geometry (CSG) 157,
    159
contaminant plume 13, 150, 152, 123
contaminants 37 et seq.
continuity 23
continuity constraint 22
contour line segment 21
contour spacing 85
contour threading 22
contours 2, 25-6, 32, 40, 48, 56, 84-5,
    115, 152, 162

convergence 106
Coombs patches 136
coordinates, Gauss-Krüger 56
coordinates, X,Y,Z 12, 137, 150
covariance 13
crest slope 66
crest spur 66
cross variogram 44
CRT 1, 84
crust 137 et seq.
CSG 157, 159
cuboids 15, 160, 167
cuesta 65 et seq.
curvature 54 et seq.

Dakota sandstone 2
DASP 59, 62
data 12
    data editing 12, 151
    data storage 13, 15, 18
    data validation 12
data models 12
data structures 1, 8, 13, 16, 58, 82, 149,
    161, 167, 181
database query 11-17, 151, 156
Dataglove 17
DCF 169, 172
dead ground analysis 93, 110
Delaunay triangles 25-6
DEM 3, 51, 56, 58, 115
depositional environment 3, 13, 83
depositional history 3, 119-20, 130
depth cues 17, 87, 105 et seq.
depth migration 130
depth sort algorithm 88
DGRM 56 et seq.
Digital Line Graph (DLG) 139
digital elevation model (DEM) 3, 51, 56,
    58, 115
digital terrain model (DTM) 81 et seq.
dimension 17
diorite 139
DLG 139
dolomite 3, 4
double connectivity filling (DCF) 169, 172
drainage basin 51
drainage divide 65
DRG 58 et seq.
DTM 81 et seq.
Dynamic Graphics 3, 17, 116, 139

Earthquake 5
economics 18
edge 15, 23, 88, 92, 134, 153, 157-60
EEC 80
ellipsoids 165
engineering 11, 80, 83, 91, 93, 99, 119
enhanced oil recovery (EOR) 115
EOR 115
erosional truncation 133
errors 12, 31, 42, 47, 56, 82, 92, 110,
    118
Euclidian space 156
Euler operators 157
European Economic Community (EEC) 80
evolving systems 13
expert systems 99, 122
exponential model 42
extrapolation 8, 141, 143
extreme values 48

Face 159
facies 12
    facies, hydro- 122-3
    facies, lithofacies 3, 12, 122, 130
    facies, seismic 129
faking data 95
faulting 3, 31-2, 130, 152
    faulting, detachment 139
    faulting, reverse 13
    faulting, strike-slip 139
    faulting, thrust 139
Felberg 64
fence diagrams 4, 6, 129, 131, 146
filtering 92, 143 et seq.
finite difference model 117, 119, 121, 125
finite element analysis 123, 125
flight simulation 82, 94
flood frequency 34, 41, 83
flood plain 37, 41
flooding simulation 58, 72
flow model 117 et seq., 151
folding 13, 136
footslope 70-2
form elements 54 et seq.
form facets 54 et seq.
form/ process relationships 74
FORTRAN 110
Fourier analysis 81, 92
fractals 81, 83-4
frame buffer 88

functionality 14, 16

Gabbro 139
Gauss-Krüger coordinates 56
Gehrden 64
generalisation 17, 82
geo-object 11
geo-relational 15
geographical information systems (GIS)
    21, 34, 51-2, 81, 85, 94, 115-6, 126,
    139, 149, 155
Geokernal 15
geologic maps 3, 139-40
geologic structure 13, 134, 140, 143, 152
Geological Survey of L. Saxony 62
geometric complexity 157
geometric transformation 87
geometrical attribute 15, 54, 62
geometrical structuring 15, 85, 129, 155
geomorphological information systems 51-
    2
geomorphological map, W. Germany 51
geomorphological mapping 51 et seq., 83
geomorphometry 51
Geopak 17
georelief 52
geoscience data 11
Geoscientific Mapping and Modeling
    System 11 et seq.
Geoscientific Resource Management
    System 155 et seq.
geostatistics 37 et seq., 125, 151
German basic map 56
Geul, river 37 et seq.
GINO 17
GIS   21, 34, 51-2, 81, 85, 94, 115-6,
    126, 139, 149, 155
GKS 62, 111
Glasgow, City of 94
Global Geoscience Transect 137 et seq.
globe 101-2
Glorious Revolution 21
GMMS 11 et seq.
good practice 12
Gouraud shading 89
gradient 54 et seq., 85, 92, 104-5
graph theory 134
Graphical Kernal System (GKS) 62, 111
gravity data 3, 137, 129, 141
Grenvillian Basement 138 et seq.

grid cell spacing 22
grid node 24
GRMS 155 et seq.
grouping factor 167
gullies 58

Hazardous waste sites 126
heavy metals 37, 39
height errors 58, 82
hermitian polynomial 28
hexahedrons 165
hidden edge removal 88, 106, 156, 159,
    165
hidden surface removal 81, 106, 110,
    111, 156, 159, 165
high angle reflection 141
hillslope 70-2
holistic modeling 17
holograms 6
hydraulic conductivity 117, 122
hydrofacies 122-3
hydrogeology 115 et seq.
hydrology 60, 123
hypsometry 64

Image processing (see also remote
    sensing) 1
information analysis 100, 107
Interactive Surface Modeling (ISM) 3, 6,
    17, 137, 139, 146-7
Interactive Volume Modeling (IVM) 15-
    16, 149 et seq.
interconectedness ratio 120
interface 8, 62, 94, 123, 181
Intergraph 17
International Standards Organisation (ISO)
    111
interpolation 8, 15, 21 et seq., 37, 40, 43,
    48, 56, 81-2, 89-90, 125, 141, 152
intervisibility 93
inverse plume analysis 123
ISM 3, 6, 17, 137, 139, 146-7
ISO 111
iso-surface 15, 152
isometric models (see also 'perspective
    view' and 'block' models) 1, 87, 115
isotropism 24
IVM 15-16, 149 et seq.

JKMAP 5

Kalman filtering 125
Kearsage-Cent. Maine synclinorium 139, 146
knowledge base 13
Kriging 13, 37 et seq.

Landform classification 51 et seq.
landscape planning 93
lease tract 152
least squares fitting 43
Leibnitz 21
Lexington Batholith 139, 146
Limburg, south 37 et seq.
limestone 3, 4, 139
linear boundaries 31
linear octree 155 et seq.
linear regression 38, 46
liquid crystal shutters 112
lithofacies 3, 12, 122, 130
log decay function 37

Mach band interference 89
magnetic data 137 et seq.
Maine 137 et seq.
Maine, Gulf of 139 et seq.
mainframe computer 11
map design 100, 111
map object 32
map-to-read 107 et seq.
map-to-see 107 et seq.
Markov chain analysis 120, 122, 125
Massachusetts Institute of Technology (MIT) 17
Media Laboratory 17
Megatek Whizzard 6
mental map 101
mesoform association 54
mesorelief 58
meta-data 13
meteorology 149
micro-computer 11, 116
microform association 54
microrelief 58
migration 140, 145
military applications 92
mini-computer 11, 95
minimum tension 151
minimum variance 45
mining 11, 37, 82, 93, 151, 155 et seq.
MOHO 137 et seq.

Morgan Hill, California 5, 7
Morton order 15
moveout techniques 141
MOVIE.BYU 160 et seq.

Nerve surfaces 136
Netherlands 37 et seq.
Newton 21-2, 24
nugget variance 43, 44

Oakwood salt dome 117
object definition language 156, 160, 165-6
object-orientation 15
octant 167 et seq.
octant search 22
octree 15, 155 et seq.
octree difference 178
octree intersection 178
octree join 178
octree union 178
oil exploration 11, 22, 119, 151
ore grades 12, 155 et seq.
orthophoto 56, 81
outcrop 4, 138
overlap 105, 106
overlays 58, 60, 62, 64, 94

Panacea 82
parabolic boundaries 32
parallax 102, 106
parallelpiped 165
parameter crisis 118
PCB 149 et seq.
pelites 139
perception 100
permanent map 101
permeability zones 118
perpendicular bisector 25
perspective view models (see also isometric models and block diagrams) 2, 6, 33, 84, 87-8, 92-3, 105-6, 149, 156
petrochemical data 129
PHIGS 110
Phong shading 89
photogrammetry 81-2
photomontage 87
Pixar Image Computer 8, 147
plan curvature 54 et seq.
plutons 139
pointers 167

pollution 37 et seq.
polygons 24 et seq., 58, 70, 80, 88-9, 94, 149, 152, 161
polyhedral topology 135
polyhedron 161
polylines 15
polynomial 22-3, 26
polytree 15
porosity 123, 152
positive feedback 13
Precambrian craton 138
price/ performance ratio (of computers) 18, 80, 99
primitive instancing 157, 158
prism maps 100, 103
probability 12
process/ form relationships 74
profile curvature 54 et seq.
Programmers Hierarchical Interactive Graphics System (PHIGS) 110
project management 8, 12, 17-18, 51, 93, 115, 137

Quadtree 15
Quebec 139 et seq.

Radius of curvature 66 et seq.
radius of search 22
raster structuring 15
ray tracing 90, 141
recomputation 15
reflection 141
refraction 141
region number 167, 169, 179
regionalisation 12
regionalised variables 37
relational model 117
relative elevation 38
relief forms 54
relief modelling 51 et seq.
relief shading 92
remote sensing 92
rendering 8, 17, 85 et seq., 181
Renderman 17
representation 1, 12, 15-17, 33, 62, 79, 81, 101, 102, 107, 155 et seq.
reservoirs 13, 118, 120
resolution 15, 58, 74, 84, 92, 108, 157, 160, 167, 179
resolution complexity 157

retinal disparity 106
Reyes rendering 86
road/ traffic engineering 93
rotation 3, 16-17, 133, 153, 156, 174-5
run encoding 15
runoff 72

Sampling 12-13, 24-27, 32, 45, 81, 89, 116, 120
sampling, random 12-13
sampling, structured 12-13, 81
San Juan Basin 2, 4
sandstone 139
SAS 62
scaling (for size) 16
scanners 139
scarp 65
scene generation 84 et seq.
Scitex scanners 139
section 16
sedimentary facies 3, 112, 122, 130
seismic exploration 129 et seq., 137 et seq.
seismic facies 129
seismograms 130
seismology 4
semiology 100 et seq.
sensitivity analysis 125
shading 89, 105, 156, 159, 181
shadow effect 22
shadowing 91, 159, 165
sign system 100, 104, 107
sill 43-4
simplicial complex 15
simulation 32, 56, 58, 60, 68, 72, 80 et seq., 118, 120-2, 129, 156
slopes 22
smoothing 17, 22-4, 32, 60
soft data 13
soil erosion 39
soils 37 et seq.
spatial addressing 11 et seq.
spatial clustering 15
spatial correlation 37
spatial functions 15, 153
spatial functions 15
   adjacency 15
   AND 15
   build 15
   centre of mass 15, 179

distance 15, 179
NOT 15
OR 15
orientation 15
peel 153
shear 15
translation 15
XOR 15
spatial indexing15-16, 160
spatial map image 103 et seq.
spatial occupancy enumeration 157, 160, 181
spatial query 15, 17
spatial uniqueness 157
spike points 92
spiral access ramp 162
spline 31, 146
Springfield, Missouri 3
Statistical Analysis System (SAS) 62
stereo 1, 4, 5
stereoplotters 80-2
stereoscopic vision 102
stochastic models 83
structured vector fields 15
submatrix 58
Sun 3/60 workstation 165
surface area 16, 60, 156, 179
surface bit array 169
surface patch 21 et seq.
surface texture 91
surfaces 17
    bicubic 85, 91, 146
    equations 135
    indeterminate 32
    intersecting 17, 130-1, 152, 161
    multi-valued 13, 134, 136
    piezometric 13
sweep representation 157-8
symbolic content 79
synclinorium 139

Tacheometers 80-1
Taconic orogeny 139
TAWS 93
TELIDON 108
temporary map 101, 108
Terrain analysis workstation (TAWS) 93
terrain databases 80, 92
tetrahedral tesselation 135
Texas Gulf coast 118

TIN 25, 115
topology 15, 54, 129-33, 135-6, 157
tracking 180-1
traffic engineering 93
trend surface 23
triangulated irregular network (TIN) 25, 115
triangulation 25, 115
tunnel 16

Unambiguity 157
unconformities 130
Uniras 17
University of Heidelberg 58
University of Strathclyde 94
urban design 93-4
US Army Engineers Topographic Laboratories 93

Validation 12
valley floor 72
variogram 37 et seq., 81
vector structuring 15
vertex 15, 89, 159, 161
vertical exaggeration 4
VIDEOTEX 108
view matrix 175, 177
view plane normal (VPN) 174-7
view reference coordinate system (VRC) 174, 177
virtual map 101, 108
vision 2, 87
visual expedients 106
visual impact analysis 80, 87, 93
visualisation 11-13, 15, 17, 79 et seq., 156, 159-60, 181
volume 16
volumes of revolution 165
Voronoi regions 21 et seq.
voxel 15, 137, 147, 155 et seq.
VPN 174-7
VRC 174, 177
VSLI 95

Weighted average 21 et seq., 37, 45
well records 3, 13, 119, 129-30, 150
wicks 24-5
wide angle reflection 141-2
Wilcox-Carrizo aquifer 117, 120
wireframe 2, 84, 88, 94

workstations 17, 80, 95, 147, 153, 181

X,Y,Z coordinates 12, 137, 150

Z buffer 88
zinc 40 et seq.
zones of influence 25-6